The
benevolent
bee

The benevolent bee

Capture the bounty of
the Hive through Science,
History, Home Remedies,
and Craft

STEPHANIE BRUNEAU

QUARRY

Brimming with creative inspiration, how-to projects, and useful information to enrich your everyday life, Quarto Knows is a favorite destination for those pursuing their interests and passions. Visit our site and dig deeper with our books into your area of interest: Quarto Creates, Quarto Cooks, Quarto Homes, Quarto Lives, Quarto Drives, Quarto Explores, Quarto Gifts, or Quarto Kids.

First Published in 2017 by Quarry Books, an imprint of The Quarto Group,
100 Cummings Center, Suite 265-D, Beverly, MA 01915, USA.
T (978) 282-9590 F (978) 283-2742 QuartoKnows.com

Quarry Books titles are also available at discount for retail, wholesale, promotional, and bulk purchase. For details, contact the Special Sales Manager by email at specialsales@quarto.com or by mail at The Quarto Group, Attn: Special Sales Manager, 401 Second Avenue North, Suite 310, Minneapolis, MN 55401, USA.

10 9 8 7 6 5 4 3 2 1

ISBN: 978-1-63159-286-7

Library of Congress Cataloging-in-Publication Data is available

Design: Bülent Yüksel
Page Layout: Laia Albaladejo
Photography: Coolburns & Co., except: Shutterstock, pages 6, 10, 11, 34, 45, 85, 91, 93, 95, 100–101, 104, 105, 106, 107, 147; iStock, page 37 (bottom); Wikimedia Commons, pages 42, 58, 66, 67, 113, 124; and Stephanie Bruneau, pages 12, 13, and 37 (top).

Printed in China

Dedication

For Emile. I am so happy at home in our hive.

Contents

Introduction

Hi! I'm Stephanie, the Benevolent Bee. I'm a beekeeper, environmental educator, amateur herbalist, artist, homemaker, and mama to two junior beekeepers in training. I'm passionate about bees, natural living, and raising healthy, creative, and happy kids.

People always ask me, "Why bees?" The answer is, I don't really know! It's just part of who I am, and I feel so lucky to have discovered my passion. Just like an artist is drawn to a canvas or a musician to an instrument, I have always been drawn to the natural world, and bees in particular. When I am working with the bees I feel calm and happy, hopeful, completely centered, and at peace. Ten years ago when my husband and I set up our first backyard hive, I knew immediately that beekeeping would be part of our life forever. That first summer we set up lawn chairs at the side of the hive and spent hours just watching the bees fly in and out.

My interest in the products of the hive came later. As our one hive became two, and then four, and then more, we soon had more honey than we could handle, plus a growing stockpile of beeswax with which to experiment. I started making simple candles and quickly fell in love with the honeyed scent and golden glow of homemade beeswax at the family dining table. Then, just after my daughter Clara turned one, she came down with a horrible cough that kept the whole family up for several nights in a row. The doctor recommended a spoonful of honey before bed, citing a well-respected study demonstrating that honey is more effective at calming coughs than the main ingredient in most cough syrups, dextromethorphan. He also suggested a vapor chest rub, adding, "You keep bees, right? If you have beeswax, you could easily make your own instead of buying the petroleum-based brand from the pharmacy." The idea of using our own bee products to help heal Clara was empowering and exciting. I found recipes for an herbal chest rub online and made a batch that evening, combining our own beeswax with organic olive oil and eucalyptus, lavender, and rosemary essential oils. Rubbing it gently into Clara's chest and feet that night felt like an expression of love and caretaking that was deeply satisfying. We all slept through the night.

These days, we sell our honey, handcrafted candles, and beeswax-based body products at local farmers' markets and craft fairs, and we use bee products daily to support our family's health and wellness. We put pollen in our breakfast cereal, boosting its nutritional value immensely. Propolis throat spray and honey citrus syrups carry us safely through cold season, and the energy boost of royal jelly powers me through long runs and tired mornings alike. Ever since I discovered that I am actually highly allergic to honeybee stings, I receive weekly doses of bee venom to reduce my body's sensitivity, and I have learned a lot about the benefits of the bee's healing sting. (No, I'm not kidding! Much more about my allergy and the benefits of bee venom later . . . .)

It turns out that there is a long documented use of the products of the honeybee hive for their delicious, healthful, and practical properties, reaching back centuries and circling the globe. Honey never spoils—the honey found in the tomb of King Tut (dating from around 1330 BCE) is just as sweet and delicious, nourishing, and fresh as the honey we harvested from our bees here in Philadelphia this past spring. Similarly, the healthful benefits of hive products, appreciated for centuries in communities around the world, are just as useful today as they were in ancient Egypt. The use of bee products has withstood the test of time, and they have been used for much longer than any scientific study or trial.

In this book, we'll explore the six main hive products: propolis, pollen, honey, royal jelly, bee venom, and beeswax. Together, we will investigate how and why bees produce these products, how they've been used by humans all over the world throughout the ages, and how today's beekeepers harvest the products. I'll also share the hive-derived recipes I use for health, wellness, and nutrition, as well as simple instructions for some of our favorite bee-based and family-friendly craft projects.

My goal with this book is to share with you my abounding enthusiasm for honeybees, my ever-growing awe of the honeybee hive and the amazing products it offers, and some of the timeless and tested uses of these products. I hope that my energy will be contagious and you will feel excited and empowered to try some of these recipes yourself. I also hope that because of this book, the wellness of your family might fit a bit more firmly within your own grasp, supported by bees in your neighborhood.

Our Teaching Apiary at West Laurel Hill

The Benevolent Bee apiary is located just outside the Philadelphia city limits at the West Laurel Hill Cemetery in Bala Cynwyd, Pennsylvania. Here, we harvest the products of the honeybee hive and observe and learn from bees; and we pass this information on, teaching about bees and bee behavior to students of all ages.

West Laurel Hill is a beautiful, sustainable, progressive, and dynamic institution that we are all thrilled to be a part of (humans and bees alike). Incorporated in 1869, it is a nonprofit and nondenominational cemetery that is also an arboretum, an outdoor sculpture park, and a place of rich history. It is a great location for our bees, and an all-around lovely spot.

Welcoming our hives is just one part of the organization's efforts to be on the forefront of sustainability—West Laurel Hill is chemical and pesticide free in their lawn care and facilities management, and even employs goats to help control invasive plants.

Our apiary can be found at the northernmost edge of the property, next to the nature sanctuary and wildflower meadow.

Choosing Bee Products

If you're a beekeeper, *The Benevolent Bee* will guide you in taking full advantage of all that your hive has to offer. If you're not a beekeeper, you'll need to purchase your products of the hive before getting started with the recipes in this book. Buy products that are local (made by the bees in your region) and treatment free (produced without pesticides or other chemical treatments): they'll be better for your health and the environment.

Buy Local

Growing support for local food, including local honey and other bee products, is in part due to the realization that purchasing food that's been shipped from far away is incredibly resource intensive, using immense amounts of packaging and fossil fuels. Most honey sold in the Unites States is cheap honey imported from China, Argentina, Mexico, or Canada. In 2015, the average price of honey from these four countries was $1.96 per pound, well below the cost of production for honey in the United States (as a point of comparison, our wholesale honey price at The Benevolent Bee averages $7.50 per pound). The super low price of imported honey undercuts local beekeepers. Because there is "no money in honey," as U.S. beekeepers often say, beekeepers who want to stay in the business and also stay afloat financially turn to commercial pollination, a by-product of commercial agriculture, where beehives are shipped thousands of miles to pollinate large single-crop fields of apples, blueberries, cranberries, almonds, oranges, and other mass-produced farm crops. Because there is no natural habitat for pollinators in these high-yield agricultural environments, the farmers pay beekeepers up to $200 per hive for pollination services during the time when the crop is in flower. These cross-country journeys are incredibly hard on the bees, both because of the stress of travel and because of commercial agriculture's intensive use of pesticides, traces of which end up in the honey.

The Benevolent Bee Teaching Apiary at the West Laurel Hill Cemetery.

In sum, supporting high-quality local bee products made by small-scale beekeepers reduces pollution and saves resources.

Buy "Treatment Free"

It is very difficult to make truly organic bee products, as bees can travel many miles from their hive location to collect nectar and pollen. Bees from hives in urban or suburban environments (like ours) inevitably visit flowers in yards or parks where chemicals have been applied, and unfortunately, this means that it's hard to eliminate traces of these chemicals in our hives and hive products. There are a few organic apiaries around the world—honey from locations so remote the beekeeper can be certain the bees had no opportunity to come into contact with chemicals. For instance, Louisiana bayou beekeeper and crawfisherman Avery Allen manages his apiaries in the Atchafalaya Basin, America's largest swamp, where the bees are acres and acres away from any populated area.

Beekeepers can, however, make certain that their bees are free of pesticides inside the hive, and can package their honey in a way that retains as many of its natural and healthful qualities as possible. Look for hive products that are "treatment free." This means that the beekeeper uses no antibiotics, pesticides, or other chemicals in her hive management. Finally, when purchasing honey, it's always best to buy honey that is "raw" (unheated) and only minimally filtered, allowing the enzymes, vitamins, minerals, and other healthful benefits of the honey to remain intact. (Unlike milk and other animal products, there is no reason to pasteurize honey, as it is its own antibiotic!)

Louisiana bayou beekeeper and crawfisherman Avery Allen tends his hives.

Photo by Mary Canning, Follow the Honey

The information and recipes in this book are not intended to treat, cure, or prevent any illness or disease. If you have or suspect that you have a medical problem, consult with your physician for diagnosis and treatment. When using bee products, always test a small amount first, and watch for any allergic reactions before proceeding. Always consult your physician or health care provider before using any bee products, especially if you have a medical problem.

Photo by Mary Canning, Follow the Honey

Bee Life Stages and Jobs

Most of the bees in a beehive are small, female worker bees. They live for 3 to 6 weeks during the summer months or 3 to 6 months during the winter, when life is less demanding. Worker bees function as a team, together accomplishing all of the necessary work of the hive. The job assigned to each bee changes as she ages.

DAYS 1–3

Hive cleaning: The first job a worker bee has is to clean the cells of newly emerged bees, including her own.

DAYS 4–12

Nurse duty: Nurse bees attend to developing larvae, feeding them and checking on them. Each larva in the hive is checked on by a nurse bee almost once every minute throughout the day!

The queen's attendants: The older nurse bees also attend to the needs of the queen, feeding her royal jelly and removing her waste as needed.

DAYS 12–18

Nectar collector: House bees collect nectar and pollen from the forager bees that are coming back from the fields and flowers. They move the nectar and pollen into empty waiting cells; they work to ripen the nectar into honey and the pollen into bee bread.

Temperature control: Fanning their wings inside the hive and at the hive entrance, worker bees carefully control the temperature and humidity of the beehive. The brood nest (where the baby bees develop) is kept between 90°F and 95°F (32°C and 35°C), even in the coldest winter months.

Wax factory: Starting at 12 days old, a worker bee is capable of producing beeswax to help form the structure of the hive. The wax is exuded in the form of scales from her abdomen.

DAYS 18–21

Guard duty: Very few bees in a hive are on active guard! These bees stand at the hive entrance, looking for intruders. Bees with an unfamiliar pheromone (scent) or creatures that aren't bees are not allowed entry.

DAYS 22–42

Field days: In the last half of her life, a honeybee ventures outside of the hive. She begins by taking "orientation flights" in the area directly adjacent to the hive, and then gradually extends the distance of her journey, visiting flowers up to several miles from home in search of nectar, pollen, and water for the hive.

The birth of a bee: A new worker emerges from her cell.

Propolis

Healing Throughout History

Lining the inside of the honeybee hive is an amazing antimicrobial "bee glue" called propolis. Every tiny gap or drafty crack in the hive is sealed with propolis, which also coats the hive entrance, walls, and even the honeycomb, providing both structural support and sterilizing action. Propolis has also been called "bee penicillin," as it inhibits the growth of any bacteria, fungus, or other unwanted microbe that might thrive in the warm and humid hive environment. In fact, the word *propolis* is derived from the Greek words *pro* ("in front of," "at the entrance to") and *polis* ("community" or "city"), meaning "before the city" or "in defense of the city" (that is, the hive). Bees also use propolis to contain potential pathogens brought in by hive intruders, such as mice. The bees kill these intruders and mummify their carcasses with propolis so their decay won't degrade the hive environment.

What Is propolis?

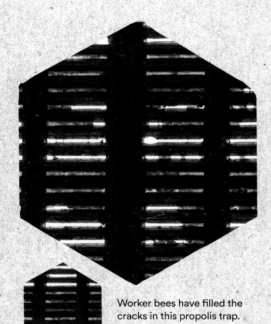

Worker bees have filled the cracks in this propolis trap.

Honeybees make propolis from tree resins that they collect from leaf buds and tree sap. Worker bees collect the resin and carry it back to the hive in the pollen baskets on their legs. Perhaps because the resin is so sticky, the worker bees cannot unload it themselves (unlike pollen); rather, they need another bee to unload their bounty for them. The bees mix the collected resin with wax, honey, and enzymes from their stomachs to turn it into the amazing and ever-useful substance that we know as propolis. The end composition is approximately 50 percent resins, 30 percent waxes, 10 percent essential oils, 5 percent pollen, and 5 percent plant debris, although each hive's propolis is a bit different, based on the variety of unique resins collected from a given hive's local trees.

Q&A

1. In recent studies, propolis has been found to inhibit the growth of antibiotic-resistant bacteria.
   a) true
   b) false

2. Propolis adds structural support to honeycomb. In fact, 1 pound of wax comb coated with propolis can hold
   a) 2 pounds (1 kg) of honey
   b) 22 pounds (10 kg) of honey
   c) 12 pounds (5.5 kg) of honey

1. ANSWER: A / 2. ANSWER: B

How Do Beekeepers Harvest
propolis?

Beekeepers are always scraping small amounts of propolis from the edges and sides of hive components, as the "bee glue" often makes maneuvering around the inside of the hive a bit of a challenge. (I keep a small glass jar in my pocket for collecting the scrapings, and after a while the small bits really add up.) To harvest larger amounts of propolis, the beekeeper places a flexible plastic screen with cracks on top of the frames in the hive, underneath the hive cover. The fastidious bees will quickly work to seal all of the cracks with propolis. This plastic screen can be easily removed and placed temporarily in the refrigerator or freezer; the propolis, which is soft and sticky in the warm hive, will quickly become brittle in the cold. Flexing the screen easily cracks off the brittle resin and it can be collected.

Collected propolis—the antimicrobial glue bees use to protect and seal the hive—after being scraped off of a beekeeper's equipment.

How Has propolis Been Used Throughout History?

"The yellow bee glue that is of a sweet scent and resembling styrax is to be chosen and which is soft and easy to spread after the fashion of mastic. It is extremely warm and attractive and is good for the drawing out of thorns and splinters. And being suffumigated [turned into smoke or vapor by applying heat from below] it doth help old coughs and being applied it doth take away the lichens [a disease of the skin]."
—Dioscorides, 40-90 CE

Propolis has been used for health and healing since ancient times—at least since the time of Aristotle (384–322 BCE), who is said to have coined the word *propolis* himself! Taking advantage of its antiseptic qualities, ancient Egyptians used propolis to embalm cadavers. In ancient Greece, Aristotle, the physician Pedanius Dioscorides (40–90 CE), and Galen (129–217 CE), a prominent Greek physician, all described medical uses of propolis.

In ancient Rome, the naturalist and author Pliny the Elder used propolis extensively. In *Natural History*, he wrote that "propolis is produced from the sweet gum of the vine or the poplar, and is of a denser consistency, the juices of flowers being added to it. Still, however, it cannot be properly termed wax, but rather the foundation of the honey-combs; by means of it all inlets are stopped up, which might, otherwise, serve for the admission of cold or other injurious influences." Pliny also wrote that propolis "has the property of extracting stings and all foreign bodies from the flesh, dispersing tumours, ripening indurations, allaying pains of the sinews, and cicatrizing ulcers of the most obstinate nature."

Some historians believe that the Hebrew word *tzori*, mentioned throughout the Hebrew Bible, was a word for propolis. *Tzori* is usually translated as a kind of fragrant medicinal resin obtained from certain kinds of trees. In Ezekiel, *tzori* is mentioned with honey twice, and in Jeremiah it is mentioned three times as having healing properties.

Propolis lies in small balls after being scraped from a beekeeper's equipment.

Propolis

Bee propolis lines the inside of a hive.

Propolis for Nutrition, Health, and Wellness

In more recent times there has been a significant amount of research on the biological activity of propolis, and many of the healing properties that so many civilizations have touted in propolis throughout the ages have been confirmed by modern-day science. More than 180 compounds in propolis have been found to have biological activity in humans, and many of them are helpful in various ways. Flavonoids—potent antioxidants and powerful antimicrobials—are the most abundant compounds in propolis. Research has demonstrated propolis's antibacterial, antifungal, antiviral, and anti-inflammatory properties and its ability to protect the liver, heal problems of the mouth and gums, and treat peptic ulcers.

Propolis can also increase the body's natural resistance to viruses and infections. In one randomized, double-blind, placebo-controlled study, 430 children were given either an herbal extract preparation or a placebo over a 12-week period in winter. The herbal preparation contained propolis (250 mg for children 1–3 years/375 mg for children 4–5), echinacea (250 mg/375 mg), and vitamin C (50 mg/75 mg). There was a 55 percent reduction in the incidence of respiratory tract infections in children taking the herbal preparation compared to the placebo, and the duration of each illness that did occur was significantly shorter (the number of days each sick child ran a fever was reduced by 62 percent)! Compelling, isn't it?

Today, propolis is used as a popular remedy. Current sales of propolis in the United States are estimated at 40,000 pounds (18,160 kg) per year. Because of its long and varied list of touted benefits, propolis is used in a range of home remedies and body care products. It is available in capsules, as an extract in alcohol or glycerin, as a mouthwash, and in many creams and cosmetics.

Philadelphia-area beekeeper and emergency room surgeon J. C. had been making home remedies from his honeybee hives for years, using them to boost his own wellness and treat personal ailments. But his respect for the healing powers of propolis grew enormously when he gave a small jar of homemade propolis cream to a woman he met with a persistent infected leg wound. He told the Benevolent Bee:

J.C.'s Story

"About three or four years ago I saw an elderly woman in my office for a sore on her leg. She had undergone a coronary artery bypass a year or two earlier and the incision on her lower leg through which they had removed a segment of her saphenous vein to do the bypass had never healed. Many things had been tried. I think she came to our practice when everyone else had given up on her. I had recently made an ethyl alcohol preparation of propolis I had harvested from my beehives. It was a pasty preparation that was dark colored and had the odor of evergreen sap. I told her I didn't have any magic bullet for her sore, but that I did have some stuff from my beehives that was known to at least help fight germs. I said it wasn't approved for anything but that if she wanted to try it I was willing to do it. So I cleaned the wound, then smeared some propolis in the wound, and let it dry before I covered it with two large bandages. (The wound itself was 2 to 3 inches [5 to 7.5 cm] in length.) I told her to try to leave the bandages on for a week if possible and then change them (sooner if necessary), cleaning out the residual propolis and reapplying some. (I gave her a test tube with some in it and some cotton swabs). Then I said come back a week after that. She did and said she was astounded that her sore had healed."

Recipes

You can easily make your own products from raw propolis. If you're not a beekeeper, a visit to your local farmers' market or beekeepers club meeting will introduce you to beekeepers in your neighborhood. Unbelievably (because it's such an amazing and useful substance), most beekeepers do not use the propolis in their hives. Most likely, a friendly inquiry will land you a small amount of propolis for free or for a reasonable price.

Raw propolis can be ground using a coffee grinder (we have one dedicated to this purpose). Ground, it can be easily infused into a topical cream or oil (for external applications), diluted in a liquid (propolis extract), or placed in capsules or mixed with honey (for ingestion). Together, these products are an amazing defense system at your service, with an ability to assist your body with healing and germ fighting.

Propolis is pretty stable, but it's still good to keep your products away from light to maximize their longevity.

Propolis Tincture

It's great to have propolis tincture on hand. At the first sign of a sore throat or upper-respiratory infection, you can add a dropperful to a glass of warm water and use it as a gargle. Or you can use the tincture as a throat spray, as described below, adding other herbal tinctures such as echinacea for even broader antibacterial and antiviral action. You can also add the tincture to body creams and salves.

Yield: Varies

Ingredients:

- *Propolis*
- *Grain alcohol (It's best to use 75 proof or higher, such as Everclear. Do not use rubbing alcohol—it is not for ingestion!)*

Directions:

> Grind the propolis in a dedicated coffee grinder.
> Combine 2 parts ground propolis to 9 parts clear grain alcohol, by weight, in a lidded glass container.
> Store in a dark place, shaking at ieast once a day for 1 to 2 weeks.
> Strain through a cheesecloth or paper coffee filter into an amber dropper bottle and store in your medicine or kitchen cabinet.

Honey Propolis Throat Spray

This powerful spray can help prevent bacterial throat infections, such as strep throat.

Yield: 6 tablespoons (100 g)

Ingredients:

- *3 tablespoons (45 ml) propolis tincture*
- *2 tablespoons (40 g) raw local honey*
- *1 tablespoon (15 ml) warm water*

Directions:

> Mix all the ingredients in a spray bottle.
> Spray in the back of the mouth any time a sore throat hits!

Propolis

Propolis-Infused Oil

Of all methods of infusion, research indicates that an oil extract of propolis may have the strongest antimicrobial effect. Applied topically, propolis oil is soothing and healing on cuts and abrasions. Propolis-infused oil can be used as an ingredient in lotions or salves (see recipes in the beeswax section); it works wonders on skin irritations or severe dryness, such as psoriasis or eczema.

Yield: 7 ounces (200 g)

Ingredients:

- *1 tablespoon (10 g) propolis scrapings*
- *7 ounces (200 g) olive, apricot kernel, or sweet almond oil (See page 143 for guidance in choosing an oil.)*

Directions:

> Grind propolis with a dedicated coffee grinder.

> Mix propolis and oil together in the top of a double boiler.

> Using a thermometer to monitor the temperature, heat oil to no higher than 122°F (50°C); higher temperatures may destroy some of the beneficial qualities contained in the propolis.

> Stir and heat for at least 30 minutes, and up to 4 hours.

> Strain the mixture through a paper coffee filter.

> The propolis that remains in the filter can be used again to make more oil; refrigerate or freeze it for another time. Store finished oil in a sealed jar in a dark place.

Herbal Mouthwash with Propolis

Because of its ability to inhibit the growth of the bacteria that cause gingivitis, gum disease, and tooth decay, propolis is a fantastic ingredient to include in oral health care products. Propolis tincture can be applied directly to a painful tooth or a cold sore for healing relief, and general oral health issues can be addressed with a mouthwash of propolis tincture in water (I recommend 1 part tincture to 9 parts water).

Respected herbalist Rosemary Gladstar has a wonderful recipe for an herbal mouthwash that does wonders keeping the dentist away. I've adapted her basic recipe to my family's tastes, and added propolis for its healing properties.

Yield: 1 cup (240 ml)

Ingredients:

- ¾ cup (180 ml) water
- ¼ cup (60 ml) vodka
- 2 dropperfuls calendula tincture
- 2 dropperfuls echinacea root tincture
- 1 dropperful myrrh tincture
- 2 dropperfuls propolis tincture
- 1 drop peppermint or spearmint essential oil (I actually like it much better without this! But others who are used to a minty flavor enjoy this optional addition.)

Directions:

› Mix all ingredients together in a glass jar.
› Swish a mouthful for about 30 seconds each night after brushing. Do not swallow.

Bee Pollen

A Nutritional Powerhouse

"Look Mama! The bees are wearing blue leg warmers today!" Four-year-old Clara is excitedly watching forager bees entering our living room observation hive. The bees are climbing up the clear plastic tubing that leads through the living room window and up into their Plexiglas home. Their little hind legs are laden with balls of purply blue pollen, very different from the shades of yellow and orange that we usually see. Where is this pollen coming from?

Walking the dog around the neighborhood that afternoon, we carefully inspect the flowering plants we see. It doesn't take long before we find the source of the blue pollen in the beautiful blue spring perennial wood squill (*Scilla siberica*). We tap the pollen grains from the flower into our open palms and appreciate their unusual royal color. Once again, I marvel at the way our bees help us open our eyes to the natural world in ways I never expected.

What is Pollen?

Pollen is a flower's way of making more flowers. Grains of pollen are a plant's male sex cells, and wind or insects carry the grains from one flower to another, ensuring genetic diversity. Certain flowering plants have developed in a way that's attractive and beneficial to bees, in a process of coevolution. The plants provide nutrient-rich pollen and sweet nectar, and in turn, the bees' hairy bodies and behaviors efficiently transfer pollen from flower to flower.

When forager bees leave the hive to collect pollen, they are very faithful to one type of flower on each visit, never visiting different flowering plants on the same trip. This evolutionary adaptation means that bees will bring apple blossom pollen to another apple blossom, where it is needed for pollination, and not a rose, where it will do no good. It also means that each pollen granule that bees bring back to the hive is uniform in color. Looking at a collection of pollen granules reminds me of looking closely at sand. From a distance, the varied colors and textures of the grains blend together, but close up it's easy to see a wide range of colors.

Pollen is important for plants, but it's also crucial for bees. Pollen is the source of all protein, fat, vitamins, and trace elements in the honeybee's diet, and it is fed to both the adult bees in the hive as well as the developing larvae. Because the queen can lay up to 1,500 eggs a day in the early summer, there can be as many as thirty thousand (!) developing larvae requiring protein to grow into bees. In one year, a hive of bees consumes about 75 pounds (34 kg) of pollen. It takes approximately one million collection trips out of the hive to gather this annual supply.

Bees collect pollen on separate trips from nectar-collecting excursions, visiting flowers that have the most nutritious and easiest to collect pollens, which can be different from the best flowers for collecting nectar. On each trip from the hive, a forager bee collecting pollen will visit between ten and one hundred flowers, and will make up to twenty trips per day. With forelegs moistened with saliva, she combs the pollen that collects on her fuzzy body and pushes it into what beekeepers call "pollen baskets," tiny Clara calls "leg warmers," and scientists call "corbicula"—a tiny cavity on the outside of each hind leg. The saliva and a bit of nectar moisten the pollen grains and help them keep together, and the tiny hairs on the bee's leg help hold the granule in place.

Back in the hive, house bees help the foragers unload the pollen into cells of honeycomb, usually in a semicircular ring just outside of an area of developing larvae, and just inside of a similar ring of honey stores. These house bees further mix the pollen with more saliva (containing enzymes from the bee's stomach) and nectar. These added ingredients start breaking down and fermenting the pollen, making it digestible for the bees. At this stage, the pollen has been turned into the super nutritious, probiotic staple of the honeybee's diet, "bee bread."

Q&A

1. One pollen granule on a honeybee's leg contains
a) 1,000–10,000 pollen grains
b) 10,000–80,000 pollen grains
c) 100,000–5,000,000 pollen grains

2. Pollen is the source of protein in a honeybee's diet. Each year, an average colony of bees consumes how many pounds (kg) of pollen?
a) 15 pounds (6.8 kg)
b) 75 pounds (34 kg)
c) 120 pounds (54.5 kg)

1. ANSWER C / 2. ANSWER B

Bee Pollen

How Do Beekeepers Harvest *pollen?*

Bees will forage for as much pollen as they need for their own nutritional requirements, which means that if a beekeeper wants to collect a small portion of the pollen for her own use or for sale, it will not harm the bees' supply—they will just ramp up their collection efforts to cover the loss.

To collect pollen, beekeepers put a "pollen trap" at the entrance to the hive. This is a special screen with holes just big enough for the forager bees to enter, but not big enough for their pollen-laden legs. The pollen granules get bumped as the bee enters the hive, and fall off into a collection tray. About half of the returning bees will enter the hive through the trap and lose their pollen. The other half will enter the hive underneath the trap, where the main entrance to the hive has been left open for business as usual. Beekeepers only collect pollen from strong and healthy hives, and only during times when pollen abounds in the bees' environment. Pollen traps are never left on for more than a week or two at a time, limiting any added stress on the hive. Beekeepers remove the pollen from the trap daily and store the collected granules in an airtight container in the refrigerator, treating the collection as a perishable product.

How Has pollen Been Used Throughout History?

Humans have been eating bee pollen just as long as they've been eating honey, as all unprocessed honey contains pollen. The unique taste and nutritional benefits of pollen are a part of what makes raw honey delicious and healthful. Allthough the history of human use of pollen is much shorter than that of honey, it has undoubtedly been part of human appreciation for the golden sweetener since the Stone Age.

The earliest mentions of pollen as a nutritional or medicinal supplement are found in books by physicians in Islamic Spain (711–1492). Also in the Middle Ages, the preeminent Jewish philosopher, astronomer, and physician Maimonides (1135–1204) recommended the use of pollen as an astringent and sedative tonic. Islamic physician, scientist, botanist, and pharmacist Ibn al-Baitar (1197–1248) promoted the use of pollen in his largest and most widely read book, *Compendium on Simple Medicaments and Foods*, touting the use of pollen as an aphrodisiac, with additional benefits for the stomach, bowels, and heart.

Palynology
The Study of Pollen

Each plant's pollen is incredibly distinct under a microscope, and different regions of the world (or even smaller areas, such as a particular meadow) have unique pollen "signatures." Pollen in a sample of honey can provide us with a relatively exact date and place for the source of the honey. In addition to pollen's use throughout history, it has been used to help us *understand* history—both ancient and recent. Scientists can use both fossilized and recent pollen to help us understand the history of plants on Earth, climactic and environmental change, and more. There is even a special field of forensic palynology that uses pollen to help solve crimes. Pollen can be used to solve cases of arson, art forgery, counterfeit medicine, and even murder, where pollen samples taken from a victim's nasal passages and clothing help police identify where the crime occurred.

Pollen for Nutrition, Health, and Wellness

C. Leigh Broadhurst, an expert in environmental remediation and trace elements in the environment and food supply, said, "All of the health benefits you've ever heard attributed to plant foods—from blueberries to broccoli to garlic—can be contained in a single pollen basket."

Indeed, Broadhurst is right. It is a bit of an understatement to simply say that pollen is highly nutritious. Pollen is full of an alphabet of vitamins and every trace element known to be essential for mammals (although some are present in very small amounts). Pollen also contains eleven enzymes and coenzymes; fourteen fatty acids; and several potent phytochemicals (plant-derived chemicals), specifically carotenoids and phenolics such as flavonoids and phytosterols, valued for their antioxidant properties. Pollen contains 12 to 40 percent protein by weight (depending on the pollen source), and unlike most plant-based proteins, it has a full spectrum of amino acids, making its protein balanced and complete.

Chock-full of these nutritious and health-boosting elements, pollen has been found to be helpful in so many ways:

- Athletes (and tired parents) can use pollen to boost energy and stamina.

- The flavonoids in pollen can help lower cholesterol and reduce inflammation, helping prevent heart disease.

- Strongly antioxidant, pollen can help neutralize the free radicals that damage the cells in our bodies.

- The wide range of nutrients and trace elements found in pollen helps boost overall health and immunity.

- Pollen contains the phytochemical rutin, which strengthens blood vessels and capillaries. This can help with varicose veins, hemorrhoids, and hypertension.

- Several scientific studies have shown pollen to be useful in treating chronic prostatitis (inflammation of the prostate gland).

- Pollen contains lecithin, which is used medically for treating high cholesterol, anxiety, and memory disorders such as dementia and Alzheimer's disease. Lecithin is also used to help control body weight. Eaten before a meal, pollen can help boost metabolism and minimize food cravings.

Nutrients in Bee Pollen

Vitamins

Biotin
Choline
Inositol
Folic acid
Provitamin A (beta-carotene)
Rutin
Vitamin B_1 (thiamine)
Vitamin B_2 (riboflavin)
Vitamin B_3 (niacin)
Vitamin B_5 (pantothenic acid)
Vitamin B_6 (pyridoxine)
Vitamin B_{12} (cyanocobalamin)
Vitamin C (ascorbic acid)
Vitamin D
Vitamin E
Vitamin K
Vitamin PP (nicotinamide)

Trace Elements

Boron
Calcium
Chlorine
Copper
Iodine
Iron
Magnesium and sodium (electrolytes)
Manganese
Molybdenum
Phosphorus
Potassium
Selenium
Silica
Sulfur
Titanium
Zinc

Bee Pollen and Allergies

Plants that rely on the wind to disperse their pollen grains have to produce pollen in the billions, hoping that a tiny portion of these grains will by chance land inside a flower on another plant of their species. For some people, that pollen is treated as a foreign protein in the body, triggering a reaction similar to one the body might have if it were encountering foreign bacteria. Sneezing, itchy eyes, runny nose—it feels like a bad cold. No fun!

The pollen that bees collect is not the same pollen that fills the air during hay fever season. Pollen from insect-pollinated plants is sticky. These plants rely on bees and other bugs to carry their genetic material from flower to flower, and they hold on tightly to their pollen until the bugs can come and pick it up.

So why do so many sufferers of spring allergies swear by local honey and pollen as a remedy for their hay fever when it's actually a different pollen causing the sneezing? I believe that pollen works for allergies because it is so full of vitamins, minerals, and phytochemicals that it boosts the body's immune system and natural defenses. For example, pollen contains the flavonoid quercetin, which is a well-touted antihistamine.

Another possibility is that perhaps the bee pollen from an insect-pollinated plant can be helpful in treating the symptoms triggered by a wind-pollinated plant that flowers at the same time. For instance, ragweed and goldenrod plants flower at the same time in the fall. Ragweed is wind pollinated, and is the cause of many fall allergies. Goldenrod is insect pollinated—it has a very sticky pollen that is highly unlikely to cause any allergic reaction, as it does not spread in the air. However, herbalists often recommend goldenrod as a remedy for a ragweed allergy.

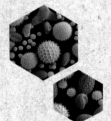

Goldenrod tea with honey is a well-known herbal remedy for countering fall allergies. Perhaps some of the helpful compounds present in bee-collected pollen grains lead to the positive effects experienced by so many people.

How to Test for a Bee Pollen

Allergy

Some people are allergic to pollen, so it's best to perform a test before adding it to your diet. Place one or two granules on your tongue and allow them to dissolve in your mouth. Wait to see if any symptoms appear. Make sure that you have allergy medication available in the event of a reaction.

Recipes

Today, bee pollen is widely available for purchase at farmers' markets, natural food stores, and online. The online mega-retailer Amazon, for example, has more than forty different varieties of bee pollen supplements for sale. It's important when purchasing bee pollen to ensure that it was refrigerated or freeze-dried at the point of harvest, ensuring that it is still fresh and potent. This is especially important if ordering online.

When you get your pollen, sample a few granules. If they're fresh, they should be soft and not crunchy or hard to chew. Each granule of pollen will taste different depending on the flower it came from—some will be sweeter, some will be more bitter—but most pollen is strong tasting, powdery, floral, and slightly sweet. Store your pollen in the fridge and it will keep for about a year.

Power Breakfast Spread

This spread is delicious on toast for breakfast, especially when topped with sliced bananas. Packed with protein, vitamins, minerals, and enough energy to power your family through the morning, it also has a taste that all ages will appreciate.

Yield: ⅔ to ¾ cup (170 to 220 g)

Ingredients:

- ¼ cup (60 g) tahini
- ⅓ to ½ cup (110 to 170 g) raw local honey
- 2 teaspoons (20 g) bee pollen
- 2 teaspoons (20 g) cocoa powder

Directions:

> Mix all the ingredients well, starting ⅓ cup (110 g) honey first, and then adding more if needed, to taste.
> Spread on toast, or eat by the spoonful for a healthy snack!

Variation: Wellness Balls

> Follow the above directions, but add any other powdered herbs or spices that you're interested in taking for their wellness benefits (such as spirulina, turmeric, maca, ashwagandha, or others). Using a teaspoon, take a scoop of the mixture and roll it into a ball on your palm (about 1 inch [2.5 cm] across).
> Place the balls on parchment paper, and store in an airtight container in the refrigerator. You can also roll the ball in shredded coconut if desired!

Honeyed Mustard and Pollen Dressing

We love to drizzle this over steamed broccoli as a nutritious side dish. It's also great over any kind of salad.

Yield: ¾ cup (180 ml)

Ingredients:

- ¼ cup (60 ml) extra virgin olive oil
- Juice of ½ lemon
- 2 tablespoons (40 g) honey
- ¼ cup (60 ml) apple cider vinegar
- 1 tablespoon (15 g) Dijon mustard
- 1 tablespoon bee pollen, ground to a powder with a mortar and pestle or coffee grinder
- 1 dash sea salt
- 1 dash pepper

Directions:

> Combine all ingredients in a blender and process to a smooth consistency.
> Store in the refrigerator, where it should last for a few months.

Morning Oats with Coconut Oil and Pollen

When we eat cereal from a box for breakfast, I can feel everyone's energy plummeting and tummies grumbling again well before lunchtime. But when we eat these hearty oats, we are all set for hours. I especially like adding the coconut oil, which in addition to tasting great has gotten lots of press lately for its health-boosting benefits and can help keep your energy high and bellies full all morning long.

Yield: 2 servings

Ingredients:

- 1 cup (80 g) rolled oats
- 1 tablespoon (15 g) raw unrefined coconut oil
- 1 or 2 dashes cinnamon
- 1 tablespoon (10 g) dried fruit (we like goji berries, but raisins work, too)
- 1 cup (240 ml) milk or milk substitute
- 1 cup (240 ml) water
- 1 teaspoon pollen
- Drizzle of honey (this is especially good with Cinnamon Honey, page 77)

Directions:

> In a medium saucepan, combine oats, coconut oil, cinnamon, dried fruit, milk, and water.
> Bring to a boil, then reduce the heat and simmer until the liquid has been absorbed and the cereal has reached your desired thickness, about 5 minutes.
> Top with pollen and honey and serve.

Bee Pollen

Afternoon Buzz Smoothie

This smoothie is great as a mid-afternoon snack. You can make it in the morning and keep it in the fridge for a boost when you start feeling a slump.

Yield: 2 or 3 servings

Ingredients:

- *1 banana, peeled (may be frozen)*
- *1 cup (245 g) mango or pineapple chunks (may be frozen)*
- *1 cup (240 ml) almond milk*
- *4 or 5 pieces crystallized ginger*
- *1 tablespoon (15 g) almond butter (or other nut butter)*
- *1 teaspoon bee pollen*

Directions:

› Combine all ingredients in a blender and process until smooth.

Variation:

› This is a great recipe to get creative with. If you'd like it to be a bit sweeter, you can add a few pitted dates. If you'd like to drink some veggies, you can add some chopped kale or spinach leaves (this recipe can accommodate up to 2 cups [35 g]). If you are not a ginger fan, skip the ginger, of course.

Honey

An Ancient Superfood

Just like Winnie-the-Pooh and his "hunny pot," you can often find members of our family with a spoon or a sticky finger in a jar of honey. We bring honey back from everywhere we travel, and the shelf above our kitchen sink has at least fifteen to twenty open jars of honey of different types at any given time. Four-year-old Clara has been known to fake a cough before bed, as she knows that this will get her a spoonful of nature's best cough syrup!

Our appreciation for the golden bounty starts with its sweet taste on the tongue, but runs much deeper, too. With amazing nutritional and medicinal benefits, a history of use spanning centuries and continents, and an immense amount of bee effort represented in every teaspoon, honey never ceases to amaze us.

A bee sips nectar from a hydrangea flower.

How Do Bees Make *honey?*

As soon as daytime temperatures rise above 55°F (13°C) and flowers are in bloom, forager honeybees are flying from flower to flower, collecting nectar to bring back to the hive to turn into honey. Whereas pollen provides the protein in a honeybee's diet, honey is the primary source of carbohydrates for the hive, fueling the tireless efforts of the industrious workers.

When a worker bee is about 22 days old, she is finally allowed to leave the hive, graduating to "forager" status. Experienced foragers will take the "newbies" on a training flight around the hive on a sunny afternoon. Forager bees are responsible for collecting nectar, pollen, propolis, and water for the hive. A forager might visit 50 to 100 flowers on every trip out of the hive, and up to 2,000 per day! This unflagging work ethic comes at a cost: bees in the warmer months only live between 3 and 6 weeks, whereas bees in the colder months can live for 3 to 6 months.

At each flower, the forager bee sucks up nectar into her honey stomach, where it mixes with enzymes and bacteria in her gut. Back at the hive, the forager transfers the nectar (proboscis to proboscis) to the honey stomach of a younger "house bee," who then transfers the nectar to an empty cell. The house bees work to evaporate the excess water in the nectar by fanning their wings, as well as sucking the nectar into their mouths and regurgitating it as many as 200 times, until it is greater than 80 percent sugar, and the thick, viscous substance we know as honey.

Once the honey is cured, the bees cap the cell with a thin layer of wax, sealing it closed. In this final state, honey can last forever without spoiling! In fact, honey found in ancient Egyptian tombs is just as stable and edible as the honey we extracted from our hives last month.

Q&A

1. How much honey does a bee make in her lifetime?
a) 1 tablespoon (20 g)
b) 2 tablespoons (40 g)
c) ½ teaspoon (7 g)

2. One pound (454 g) of honey is made from the nectar of . . .
a) 2,000 flowers
b) 2 million flowers
c) 900,000 flowers

1. ANSWER: C / 2. ANSWER: B

In its final state, honey can last forever without spoiling.

Bees engage in the sweetest of exchanges: transferring honey from one proboscis to another.

Honey from the manuka flower is prized for its antibacterial properties.

What is honey?

Just as lavender flowers smell different from rose, geranium, or hyacinth, honey made from the nectar of different flowers tastes different, too, and has different properties. Honey made from the nectar of buckwheat flowers is very dark in color, has a deep molasses-like flavor, and is strongly antioxidant. Orange blossom honey is lightly sweet, with a hint of citrus. Manuka honey, from New Zealand's manuka (or tea tree) bush, is highly prized for its antibacterial properties. The nectar from the manuka flower contains methylglyoxal, an anti-bacterial organic compound that is transferred to the honey. When you purchase manuka honey you will notice "UMF" (Unique Manuka Factor) ratings on the jar, and jars with a higher UMF rating are more expensive. The ratings indicate the amount of methylglyoxal present in the honey. Honey with a UMF rating greater than 10 is considered to be highly therapeutic, and is used around the world to treat wounds and burns.

If a beekeeper wants to collect a true varietal honey, she waits until a flower bloom from the desired nectar source has just begun, moves her hives to that location, and then harvests the cured honey after the bloom is over. At The Benevolent Bee, we don't move our hives, and our honey is true Philadelphia honey—collected from flowers anywhere from 1 to 6 miles (1.6 to 9.7 km) from our hives' location. This includes front and back yards, city street trees, community gardens, arboretums, and more. The taste of our honey changes both seasonally and annually, because weather conditions change from year to year and the composition of blooms at any given time is always unique.

Benevolent Bee honeys:
same hive, different harvests.

In addition to nectar, raw honey contains pollen and other incidental grains from hive life, little bits of wax and propolis from the honey extraction process, as well as enzymes and bacteria transferred to the honey from the bees' stomachs. Modern processing techniques often involve super-filtering honey for clarity and superheating it to avoid crystallization. These processes diminish the healthful qualities of honey, removing the residual pollen and destroying the enzymes, bacteria, and nutrients. Look for honey that is labeled "raw"—this label isn't regulated, but generally it means that the honey is strained (run through a screen to remove large particulate matter like chunks of beeswax) rather than fine-filtered, and not heated above ambient temperatures that could occur within the hive (generally nothing greater than 100°F [38°C]).

How Do Beekeepers Harvest honey?

Beekeepers harvest honey from their beehives at the end of a major nectar flow, always making sure to leave enough honey in the hive to satisfy the needs of the bees. This is especially important when going into fall and winter, as a healthy beehive requires *at least* 60 pounds (27 kg) of honey to make it through the winter. In our apiary, we try to leave each hive with 80 to 100 pounds (36 to 45 kg) of honey, as we'd rather leave too much than too little, risking the hive's starvation.

When a beekeeper wants to harvest honey, she takes a frame of capped honey from the hive and slices the wax cappings off with a hot knife, revealing the honey stored in the cells of the comb. Uncapped frames are put into a honey extractor—a stainless steel drum that holds the frames—and spins, flinging the honey out of the cells using centrifugal force. The wax comb stays intact and can be reused by the bees. Occasionally we will choose to crush the comb and strain the honey from the crushed wax. This is more energy intensive for the bees, who then have to recreate the wax, but it yields a larger wax harvest for the beekeeper and keeps the comb in the hive fresh, avoiding potential pesticide build-up in the hive.

Stephanie and Emile hold two frames of capped honey ready for extraction.

We use a hot electric knife to slice off the wax cappings.

The uncapped frame is now ready for the extractor.

Clara spins the frames in the extractor.

The honey in the comb is drawn out through centrifugal force.

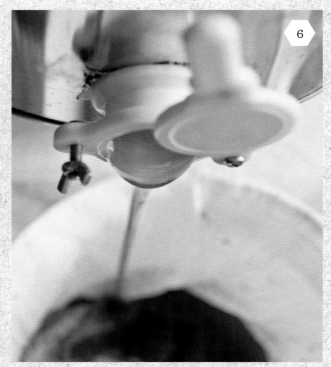

A spout at the bottom of the extractor allows the honey to run through a simple mesh filter before being placed into jars.

How Has honey
Been Used Throughout History?

Humans have been enjoying honey since the Stone Age; the oldest-known depiction of human collection of honey is a prehistoric cave painting in Valencia, Spain, thought to be at least 8,000 years old, predating modern civilization. Documentation of the use of honey is found in almost all cultures and many historical texts since the beginning of human society.

The earliest evidence of organized honey production is found in ancient Egypt, where stone carvings dating to 2400 BCE depict bees, hives, and men harvesting honey and filling and sealing jars. Honey's use in ancient Egypt is well known: Egyptians used honey as currency, food, and medicine, and to embalm and to honor the dead. The Egyptian *Papyrus Ebers*, the oldest-known book of medicine (fifteenth century BCE), includes more than 800 medicinal recipes, many of which employ honey for preventing and curing infection.

A honey hunter at work, as depicted on an 8,000-year-old cave painting near Valencia, Spain.

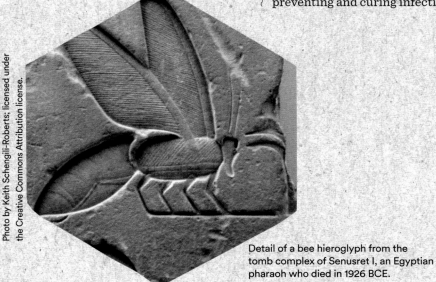

Detail of a bee hieroglyph from the tomb complex of Senusret I, an Egyptian pharaoh who died in 1926 BCE.

The first recorded medicinal use of honey dates to between 2100 and 2000 BCE in Sumer (what is now Iraq). An unearthed Sumerian clay tablet provides instructions for preparing what appears to be some kind of salve for external application: "Grind to a powder river dust . . . and . . . then knead it in water and honey, and let . . . oil and hot cedar oil be spread over it."

A page from the Ebers Papyrus, one of the oldest medical texts in the world; many of the recipes call for honey.

In the Ayurvedic tradition of ancient India, honey was used to treat infected wounds, and was mixed with ghee (clarified butter) to make a post-surgical paste (c. 1400 BCE). Cleopatra (69–30 BCE, Egypt) famously bathed in milk and honey, and she compiled a book of grooming tips, dedicating many pages to the use of honey. More recently, Russians and Germans used honey in World War I as an antiseptic to prevent wound infection, and combined cod liver oil and honey to treat burns and other wounds.

From Hippocrates (469–399 BCE) to Jennifer Lopez (born 1969), who credits her clear and honeyed complexion to her use of a honey and lemon face mask, human use of honey spans continents and centuries.

Honey

"[Honey] causes heat, cleans sores and ulcers, softens hard ulcers of the lips, heals carbuncles and running sores."

Hippocrates (469–399 BCE, ancient Greece)

"[Honey] will clear freckles from the face in a trice, of this about three ounces may suffice."

—Ovid (43 BCE–18 CE, Roman poet)

"[If] the dew is warmed by the rays of the sun, not honey but drugs are produced, heavenly gifts for the eyes, for ulcers and the internal organs."

Pliny the Elder, in *Natural History* (23–79 CE, ancient Rome)

"Honey is a remedy for every illness...."

—The Prophet Muhammad (570–632 CE)

"There proceedeth from their bellies a liquor of various color, wherein is medicine for man."

—Qur'an 16:69 (609–632 CE)

"[Honey] helps to kill pain and relieves internal heat and fever and is useful to many diseases. It may be mixed with many herbal medicines. If taken regularly, one's memory may be improved, good health ensues, and one may feel neither too hungry nor decrepit."

—Shen Nong's *Book of Herbs* (second century BCE, China)

"My son, eat thou honey, because it is good; and the honeycomb, which is sweet to thy taste...."

Proverbs 24:13

Honey
for Nutrition, Health, and Wellness

With a shelf life of "forever" and healing qualities that seem supernatural, honey has been used for ages in nutrition, health, and healing. Is it magic? Maybe! But it's also science. In more recent times, scientific study has helped us understand some of the reasons why this golden healer is so healthful and helpful.

Raw honey has been found to be antifungal, antiviral, and strongly antibacterial—so strongly antibacterial, in fact, that it can inhibit the growth of antibiotic-resistant superbugs, like MRSA (Methicillin-resistant Staphylococcus aureus). "The unique property of honey lies in its ability to fight infection on multiple levels, making it more difficult for bacteria to develop resistance," says researcher Susan M. Meschwitz, speaking about findings from her study on honey's antibiotic properties, in a presentation at the National Meeting of the American Chemical Society.

The Science Behind the Magic of Honey Healing

So how does honey work its magic? First, honey is hygroscopic, meaning it attracts and retains moisture from everything it comes into contact with, including bacteria and microbes. Researchers say you can watch this process under the microscope, cheering at the microscopic victory as honey sucks the water out of the bacteria, rendering it inert. Second, with a pH between 3.2 and 4.5, honey is slightly acidic, a quality that inhibits the growth of pathogens. Honey also contains glucose oxidase, an enzyme transferred to the honey from the bees' stomachs. Glucose oxidase releases hydrogen peroxide, which is strongly antibacterial.

Honey and Gut Health

Honey can help encourage balance in the gut. Manuka honey has been shown effective in preventing the growth of *H. pylori*, the bacteria that is responsible for most stomach ulcers and much abdominal discomfort. And in addition to discouraging bad bacteria, honey has been found to promote the growth of beneficial bacteria, such as *Bifidobacteria* and Lactobacillus, which have been shown to feed on honey as a "prebiotic;" they become more plentiful and stronger with honey as a food source. This makes honey useful for individuals struggling with a bacterial imbalance, such as thrush or irritable bowel syndrome (IBS). The enzymes in honey have also been shown to aid digestion and boost the assimilation of nutrients from the food we eat.

Honey and Burns

Used topically, honey is an effective treatment for burns, and has modern scientific backing. A study in the *Journal of Cutaneous and Aesthetic Surgery* (2011) found that treating burns with honey instead of the standard treatment of silver sulfadiazine could cut healing time almost in half (from an average of 32 days to an average of 18 days). Derma Sciences, a medical device company, has been marketing and selling a product called MEDIHONEY—bandages covered in honey—that is now used in hospitals around the world.

To use honey on burns, first run the burn under cool (but not cold) tap water immediately to reduce the temperature. Next, apply honey directly onto the burn, or soak gauze in honey and apply that to the burn. Then, wrap a secondary dressing on top of the first layer of gauze to prevent the honey from oozing out. Change the dressing once or twice daily.

Note: Serious burns require immediate medical attention. If your burn is larger than 1 inch (2.5 cm), or if you suspect you have a third-degree burn, seek medical attention and do not attempt to treat it on your own.

If you want to give yourself a honey facial, first, if you have long hair, tie it back! Then, using the back of a spoon, apply raw honey to your face, avoiding the eye area, by simply smoothing a thin layer across your skin. Let it sit for about 10 minutes and then rinse with warm water. Enjoy softer, moister skin immediately!

Honey
and Skin Care

Another popular topical use of honey is for healthy and radiant skin. Honey is a humectant, meaning it promotes the retention of water and so holds moisture in the skin. This quality, combined with its antibacterial properties, makes honey a fantastic skin mask and cleanser. Actress Scarlett Johansson puts honey on her skin: "You just warm your face so that your pores are open . . . and then you just take a spoon and apply the honey directly to your face and leave it for 10 to 15 minutes," she told Style.com in 2011. "It really adds an amazing glow and your skin is so soft afterwards." Between twenty-first-century celebrities and Cleopatra (with her milk and honey baths), I know I'm in good and historic company when I put honey on my face at night!

Recipes

Mary Poppins uses "a spoonful of sugar" to help the medicine go down, but in our house we use a spoonful of honey. Two great ways to use honey for nutrition, health, and wellness are honey syrups and infused honeys. Honey syrups take advantage of honey's hygroscopic quality—its amazing ability to thoroughly draw the juices out of anything it comes into contact with. Mixing honey with fresh herbs and fruits will create a sweet syrup flavored by your chosen ingredients and the healthful qualities they contain. Infused honeys are made with dried herbs and spices. They take longer to make but can be a great way to boost the healing properties of your honey (and they taste great, too).

Note: Never feed honey to infants younger than one year of age. Honey can harbor botulism spores. Although the toxin does not affect adults or children with mature gastrointestinal tracts, the spores can germinate when swallowed by infants, and can be very harmful.

Turmeric Honey

Turmeric root is a beautiful, bright orange color. It has a bitter flavor that stimulates metabolism and aids in detoxification. It is excellent for the liver, stimulating the flow of bile, supporting digestion, and performing other important metabolic functions. By supporting liver function and aiding in gentle, sustainable detoxification, turmeric helps reduce heat and inflammation throughout the body. It is a great holistic tonic for those with chronic, excessive inflammation, including that associated with arthritis, rheumatism, and seasonal and environmental allergies. Regular use can also help reduce pain associated with menstruation. To support such ailments, incorporate turmeric into the diet and/or take it as a dietary supplement.

Yield: ½ cup (160 g)

Ingredients:

- *¼ cup (26 g) dried, powdered turmeric root*
- *½ cup (160 g) raw, unrefined honey*

Directions:

› Combine powdered turmeric with honey in a glass jar with a tight-fitting lid and mix with a spoon until well combined. This preparation is ready to use immediately. Store at room temperature.

› Use Turmeric Honey in cooking or take it by the spoonful as a dietary supplement (½ to 1 teaspoon, 1 or 2 times a day).

Cinnamon Honey

Cinnamon is warming, delicious, and healthful. Cinnamon-infused honey is great in tea, spread on toast with butter, or drizzled on your morning oats or pancakes as a syrup alternative. This honey makes a lovely gift item—be sure to leave a cinnamon stick in the jar for the extra spice and aesthetic benefits!

Yield: 16 ounces (454 g)

Ingredients:

- *16 ounces (454 g) raw local honey*
- *5 sticks cinnamon, plus more to share*

Directions:

› Place honey and cinnamon sticks in a medium-size pot (running some warm water over the honey jar will make it easier to pour) and heat it uncovered over low heat.

› Make sure the honey does not come to a boil (you'll lose some of its health benefits, and it will also create a huge mess if it boils over!).

› Cook for 10 minutes, stirring occasionally.

› Let cool (but keep it warm enough that it's still easy to pour).

› Pour honey into a glass jar, add a few more cinnamon sticks, and seal.

› The longer the honey sits, the more of a cinnamon flavor it will take on.

› You can taste it periodically and remove the cinnamon sticks once the honey has reached your desired flavor.

› If you'd like it really cinnamony, you can add 1 teaspoon of ground cinnamon to the mixture as well.

Note: The Turmeric Honey recipe was contributed by our friends at Thyme Herbal in Conway, Massachusetts (thymeherbal.com). Brittany Wood Nickerson, an amazing herbalist, nutritionist, and cook, as well as the founder and primary instructor at Thyme Herbal, teaches a three-year Herbal Apprenticeship Program, as well as courses in herbal cooking and homesteading. We've learned so much from her!

Elderberry Syrup

We make this syrup a lot starting in September, when "back to school" means that the sunscreen and bug spray on our shelf are quickly replaced by immune boosters and tissues. Elderberries contain high levels of vitamins A, B, and C, and can stimulate the immune system. Combined with the healthful and healing properties of honey, this syrup is a delicious tool for combating colds and flus.

Yield: 2¾ to 3 cups (650 to 710 ml)

Ingredients:

- 1 cup (100 g) dried black elderberries
- 3½ cups (830 ml) water
- 1 teaspoon grated fresh ginger
- 1 cinnamon stick
- 2 or 3 whole cloves
- 1 cup (340 g) raw honey
- 2 ounces (60 ml) brandy (optional, for adults only)

Directions:

› Combine elderberries, water, ginger, cinnamon, and cloves in a small saucepan. Bring to a boil over high heat, and reduce to a simmer.

› Simmer uncovered for 45 minutes, until the liquid has reduced by almost half. Watch carefully so the mixture doesn't evaporate fully or boil over.

› Remove from the heat and let sit until mixture is cool enough to handle.

› Mash the berries using a spoon and pour through a strainer into a glass jar.

› Discard the elderberries and spices, and let the liquid cool to lukewarm.

› Add the honey and mix well. Store in a glass jar in the refrigerator, where it will last for a few months. If this is just for adult use (not kids too), you can add the brandy to lengthen the shelf life.

› Take by the spoonful as a daily immune booster (the standard dose is ½ to 1 teaspoon for children, and ½ to 1 tablespoon for adults, once a day if you're not sick but 3 times a day in times of illness), or add to sparkling water with a slice of lemon for a healthful and refreshing drink!

Honey

Lemon Ginger and Sage Cold and Cough Syrup

Combine honey with the healing properties in lemon, ginger, and sage, and you've got a natural, delicious way to treat cold and flu symptoms. Sage has been used in medicine in cultures around the world for centuries. The herb is antibacterial and antispasmodic, making this syrup a great choice for a cough at bedtime. This syrup can be swallowed straight from a spoon or stirred into hot water or tea.

This recipe is very flexible. Add more ingredients for larger jars or for a stronger syrup. Try substituting orange slices for the lemon, or cloves for the sage (cloves are a great choice, as they are an expectorant and antiviral!).

Yield: 1 cup (about 200 g)

Ingredients:

- *1 lemon*
- *1 teaspoon ground ginger or 2 teaspoons grated fresh ginger*
- *6 to 8 fresh sage leaves*
- *½ cup (170 g) honey (or enough to fill the jar)*

Directions:

> Slice lemon into thin half-moon shapes. Add half of the slices to a half-pint jar, then add half of the ginger and sage, then half of the honey.
> Stir with a chopstick or other long, thin object. Move citrus slices around to make sure the honey runs between them.
> Repeat with remaining ingredients, then stir and top off with more honey if necessary to fill the jar.
> Within 3 or 4 hours the honey will have drawn the juices out of the other ingredients.
> Give it a stir to combine the honey and juice. The thick honey will have transformed into a lovely syrup, which will get stronger with time.
> Store in a glass jar in the refrigerator for up to 2 months.

Note: Sage should not be used when pregnant or breastfeeding.

Honey Fire Tonic

This is a traditional folk recipe that combines apple cider vinegar and raw local honey with powerful and spicy antimicrobial and decongestant herbs to boost your immune system, kick-start your circulation, stimulate digestion, and warm you right up. I like to drink it as a tea, adding about an ounce (30 ml) of tonic to a cup of boiled water. When I'm getting sick, I drink this tea three or four times a day. You can also take it straight up by the spoonful, drizzle it on food (like steamed veggies) as a dressing, mix it with lemonade or other juices, or add it to a cocktail (like a Bloody Mary).

Yield: About 1 quart (940 ml)

Ingredients:

- ½ cup (64 g) fresh grated ginger
- ½ cup (64 g) fresh grated horseradish
- 1 onion, chopped
- 8 to 10 cloves garlic, crushed or chopped
- 1 organic hot pepper, such as jalapeño or habanero, chopped
- Zest and juice from 1 lemon
- Several sprigs fresh rosemary or 2 tablespoons (6 g) dried rosemary leaves
- Several sprigs fresh thyme or 2 tablespoons (6 g) dried thyme leaves
- 1 tablespoon (9 g) turmeric powder
- ¼ teaspoon cayenne powder
- 3 cups (710 ml) apple cider vinegar (enough to fill a quart-sized jar)
- ¼ to ½ cup (85 to 170 g) raw local honey, to taste

Directions:

- Place all of the ingredients in a lidded quart-size jar. Close tightly.
- Shake well to combine. Store in a dark, cool place for 3 to 6 weeks, shaking daily.
- After a month, use cheesecloth or a fine-mesh strainer to strain out the pulp, pouring the tonic into a clean jar.
- Store the finished tonic in a dark, cool place for up to one year.

Golden Milk

Turmeric and ginger are immune-boosting, anti-inflammatory, and beneficial to digestion. This drink is especially good after dinner and before bedtime, in place of a heavy dessert. The black pepper in this recipe is important; adding black pepper to turmeric-spiced food activates the curcumin—the active ingredient of turmeric—and increases its bioavailability enormously.

Yield: 3 cups (710 ml)

Ingredients:

- 3 cups (710 ml) milk or milk substitute (we like coconut milk best here)
- ½ inch (1.3 cm) fresh ginger, chopped
- Dash of ground cinnamon
- Pinch of ground black pepper
- 1 teaspoon ground turmeric
- 1 tablespoon (20 g) honey

Directions:

> Combine the milk, ginger, cinnamon, and pepper in a medium-size saucepot, cover, and simmer over medium-low heat for 15 minutes.

> Remove from the heat and strain out the ginger. In a small bowl, stir the turmeric into the honey, and then stir the mixture into the milk until dissolved.

> Drink to your health! Any leftover milk can be stored in the refrigerator for up to 3 days.

Royal Jelly

The Queen's Fountain of Youth

Did anyone ever tell you "you are what you eat"? There is no creature for which this is truer than the honeybee. Amazingly, queen bees are genetically exactly identical to worker bees. But they're fed a different diet from worker bees their whole lives, from the time they are tiny larvae until the day they die. This different meal plan causes their physiology and behavior to develop completely differently from worker bees, despite the same genetic foundation. What is this magic food? It's not broccoli! It's the aptly named substance royal jelly.

What Is royal jelly?

Royal jelly is a protein-rich excretion from the glands of worker bees—I think of it as a honeybee's version of mother's milk. Although all larvae are fed royal jelly for the first three days of life, larvae chosen by the worker bees to become queens are bathed in it throughout their development in special, elongated "queen cells." After the larval stage is complete and the queen bee emerges, she is fed royal jelly throughout her life.

Queen bees are one and a half times the size of worker bees, live three to four years (instead of three to six weeks), are sexually mature (unlike worker bees, who cannot mate), and have a different set of behaviors from the other bees in the hive. When a hive needs a new queen, it will select up to ten eggs less than three days old, and begin feeding the newly hatched larvae royal jelly. The first queen to emerge will sting the other developing queens through their cells, killing them before they can hatch. If two or more queens hatch at the same time, they will fight to the death!

About three to five days after emerging, on a sunny day with low wind, the new queen will take her "nuptial flight." She will find a drone congregation area—a place where male (drone) bees from other hives hang out and wait for a queen, and will mate with twelve to twenty drones in midair, gathering a much genetic material as she will need for her entire life (up to 6 million sperm!).

Back at the hive, the queen's main role is reproducer in chief. She controls the size of the hive, laying more eggs in preparation for spring and summer, and slowing laying in preparation for the cooler months, when there's less work to do and less food around. In the height of the spring, the queen can lay up to 1,500 eggs a day. This is more than her own body weight in eggs! The queen is always surrounded by a circle of devoted workers who feed her constantly and dispose of her waste. They also collect and distribute her queenly pheromones throughout the hive, letting all the hive's residents know that their queen is alive and well.

It's amazing to think that such a different set of abilities and behaviors can arise from the simple matter of food! It makes you want to take your vitamins, doesn't it? Or perhaps just a nice healthy dinner, followed by a good spoonful of royal jelly–infused raw honey.

Photo by Waugsberg; licensed under the Creative Commons Attribution license

Queen bee larvae float in a bath of royal jelly.

How Do Beekeepers Harvest  royal jelly?

Nurse bees hard at work tending to larvae in a queen bee rearing system.

It is no wonder that royal jelly is an expensive product, as its production is a painstaking process that requires close attention and precise timing. First, a beekeeper creates a small colony of bees with no queen. She ensures that this small colony has many young bees that will work as nurse bees in the hive. Next, she inserts fake queen cups into this colony (these are several rows of plastic or wax cups that are the right size for bees to build queen cells on), with each cup containing a honeybee egg, hand grafted into each cup. Instinctively, the workers will start raising queens for their colony, using the eggs and queen cups provided. The nurse bees will fill the cups with royal jelly. At the perfect moment (usually between the second and fourth day of larval development), the beekeeper will remove the royal jelly from the queen cups with a small suction tool. If she is too early or too late, there won't be enough royal jelly to harvest. Working in this way, a beekeeper can harvest about 17 ounces (500 g) of royal jelly per hive in a season.

Q&A

1. Worker bees live for 3 to 6 weeks in the summer or 3 to 6 months in the winter. Queen bees, genetically identical to worker bees but fed a diet that includes royal jelly, have an expanded life expectancy of...
a) 10 to 12 months
b) 1 to 2 years
c) 3 to 4 years

2. Worker bees are sterile; physiologically different as a result of her royal jelly diet, the queen bee is the only bee in the hive capable of mating and laying fertilized eggs. How many eggs can the queen bee lay per day?
a) 500
b) 1,500
c) 50

1. ANSWER: C / 2. ANSWER: B

How Has royal jelly Been Used Throughout History?

In cultures around the world, royal jelly has been used to promote a healthy and long life. Because it's such a precious substance (so little is made, and it's not easy to harvest), historical use was mostly for royalty, just like in the honeybee hive. In traditional Chinese medicine, royal jelly is called "food of the emperors" and has long been prescribed to lengthen life, promote energy and vitality, and prevent illness. Royal jelly has also been used by the maharajas of India, who have long valued the substance as a key to maintaining youthful energy. In ancient Egypt, royal jelly was given to the pharaohs, promoting their longevity. More recently, Pope Pius XII (1876–1958) was prescribed royal jelly from his physician to help him recover from a severe illness. Princess Diana (1961–1997) was documented to have used royal jelly throughout her pregnancy to help with morning sickness, and Queen Elizabeth (currently in her nineties) uses royal jelly regularly to "stave off fatigue."

Today, royal jelly is used by more than royalty. China is recognized as the world's largest producer and exporter of royal jelly, with an estimated annual production of between 400 and 500 tons. Almost all of the exports are to Japan, Europe, and the United States. Korea, Taiwan, and Japan are also important producers and exporters. The price is about $45 per pound (454 g) in bulk, but can be much higher in a processed form (such as tablets, capsules, or vials), where 1 pound (454 g) could cost the consumer as much as $1,500.

royal jelly
for Nutrition, Health, and Wellness

Royal jelly is a nutrient-rich liquid. It is roughly 66 percent water, 13 percent protein, 15 percent carbohydrate, 5 percent fatty acids, and 1 percent trace elements. Like bee pollen, it contains a complete and balanced set of amino acids. It contains all the B vitamins and traces of vitamin C. Royal jelly also contains collagen (the major protein in our skin, hair, nails, bones, and veins) and several antioxidant enzymes.

Today, royal jelly is sold widely as a nutritional supplement, although this is perhaps a miscategorization, as it is taken in such small amounts (250 to 500 mg at a time) that to use it as a source of dietary nutrients doesn't really make sense. Rather, it is more often used for its ability to act as a stimulant or "wellness boost."

Properties of Royal Jelly

Properties that have been attributed to the "food of the emperors" include:

· Increased oxygen consumption

· Enhanced energy and endurance

· Increased appetite

· Heightened libido and fertility

· Help during menopause and difficult menstrual periods

· Expanded resistance to viral infections such as influenza

· Stabilized blood pressure

· Treatment for anemia

· Ability to balance cholesterol levels

· Cosmetic effects for skin and hair

Royal Jelly and Stamina

Many of the royal jelly supplements sold today tout its ability to boost endurance and combat fatigue and stress. A 2001 Japanese study bolstered these claims, finding that the endurance of mice increased when they were fed fresh royal jelly.

Royal Jelly and Cholesterol

In addition to boosting energy, several studies have found royal jelly to be cardioprotective, lowering blood pressure and reducing LDL or "bad" cholesterol. A 1995 study found that just 0.1 gram of royal jelly (dry weight) decreased cholesterol by 14 percent.

Royal Jelly and Immunity

C. Leigh Broadhurst, a USDA researcher, postulates that just like mother's milk helps boost the immunity of newborn mammals, royal jelly provides this service to the immune system of a honeybee colony. Royal jelly has also been found to be highly antimicrobial, effective against yeast and bacteria. A 1995 study conducted in Egypt demonstrated that royal jelly was capable of killing several different kinds of bacteria, including *Staphylococcus aureus* and *E. coli.*

Royal Jelly and Fertility

The well-known children's author Roald Dahl wrote a short story titled "Royal Jelly," in which a beekeeper boosts his fertility and the health of the eventual baby by secretly supplementing the family's food with royal jelly from his hives. And, in fact, it's not just fiction—royal jelly's positive effects on reproduction have been scientifically validated in animal studies, perhaps because of its ability to balance hormones or boost overall wellness and nutrition. In three different studies, rabbits fed a royal jelly supplement had increased fertility, quail reached sexual maturity faster and laid more eggs, and chickens had increased egg production.

Royal Jelly and Skin Care

Royal jelly has also been used cosmetically. It boosts the production of collagen, what many think of as the key to youthful-looking skin. Wrinkles form when our skin loses its collagen. In cosmetics, royal jelly is used to promote skin elasticity, regrowth, and rejuvenation. It's also used in shampoos and conditioners, and increasingly so! Formerly a specialty product, royal jelly can now be found in shampoos sold at many common retail outlets.

There is still a lot to be learned about royal jelly. Although personal anecdotes attribute many benefits to royal jelly, there hasn't been as much scientific validation as there has been with other bee products. This could be, perhaps, because there is less available funding for high-quality research. Most of the research that does exist comes out of Asia, as it is used more regularly there than in the United States. Still, there are many personal testimonies to the benefits of royal jelly in many civilizations throughout time—from today's British royal family to Chinese emperors in centuries past.

Will You Rear Good Bees?

Unlike worker bees with their relatively short life expectancy, queen bees can live for up to four years. Many beekeepers mark each queen with a dot of paint on her thorax. This helps the beekeepers keep track of the age of the queens in their hives and identify the queens more quickly. Across the world, all beekeepers follow the same color-coding convention:

- Queens born in years ending in 1 or 6 are marked with WHITE paint.
- Queens born in years ending in 2 or 7 are marked with YELLOW paint.
- Queens born in years ending in 3 or 8 are marked with RED paint.
- Queens born in years ending in 4 or 9 are marked with GREEN paint.
- Queens born in years ending in 5 or 0 are marked with BLUE paint.

How can you easily remember this international color code? The simple mnemonic Will You Rear Good Bees makes it easy.

Recipes

Royal jelly is perishable, and purchasing a professionally processed product is recommended. The best way to purchase it is in freeze-dried capsule form, or mixed with raw honey, which acts as a natural preservative. If you do purchase it unprocessed, I recommend that you purchase it frozen (don't purchase royal jelly that has been stored or shipped at room temperature). The recipes included in this book call for royal jelly mixed with honey—we purchase a version of royal jelly in honey where every 1 teaspoon contains 675 milligrams of royal jelly.

As with all bee products, test a small sample first to make sure you're not allergic.

10-HDA: A Tip on Buying Royal Jelly

The main fatty acid in royal jelly is the 10-hydroxy-2-decenoic acid, or 10-HDA. It is only found in royal jelly and cannot be made artificially. A good-quality and fresh royal jelly should contain between 1.4 and 2.2 percent 10-HDA. When you buy royal jelly, look for the HDA content on the label; if 10-HDA levels are low or unlisted, the sample may have been adulterated or damaged, and may not be from a reputable manufacturer.

Royal Jelly Fudge

This is a delicious and healthy treat that you don't have to feel guilty about eating.

Yield: 12 pieces

Ingredients:

- *4 tablespoons (80 g) royal jelly and raw honey mix*
- *4 tablespoons (60 g) unrefined coconut oil, melted*
- *3 tablespoons (24 g) raw cacao powder*
- *2 to 3 tablespoons (30 to 45 g) nut butter*

Directions:

> Combine all the ingredients in a medium-size bowl and stir until well blended.
> Pour the mixture into a small pan lined with waxed paper, or evenly into a silicone ice cube tray.
> Freeze until solid and cut into 12 even pieces.
> Store in the freezer to keep solid.

Royal Jelly and Honey Mint Tea

~~~~~~

This is a variation of a recipe created by an apitherapist to help normalize blood pressure. With the caffeine in the green tea and the energy boost from the royal jelly and raw honey, this refreshing drink should kick-start your day or give you a nice afternoon boost. I like it best iced.

Yield: 4 cups (940 ml)

Ingredients:

- *4 cups (940 ml) water*
- *2 bags organic green tea*
- *A big handful of fresh mint leaves*
- *1 tablespoon (20 g) royal jelly and raw honey mix*
- *Freshly squeezed lemon or lime juice*

Directions:

› Bring the water, tea bags, and mint to a boil in a medium-size saucepan.

› Simmer for 5 to 10 minutes or until the desired minty flavor is achieved.

› Remove from the heat and strain out the mint leaves and tea bags. Stir in the royal jelly honey.

› Add lemon or lime juice to taste. Serve hot, or chill in the refrigerator and serve over ice (with a citrus wedge and a sprig of mint if you're feeling fancy!).

Fertility Smoothie

Royal jelly balances hormones and has estrogenic effects in the body, which can help with fertility. Similar to royal jelly, maca (the root of a plant grown at elevation in the Peruvian Andes) is also rich in minerals and nutrients, boosts libido, and has been known to help with hormone balance and fertility.

Yield: 1 or 2 servings

Ingredients:

- *1 cup (255 g) frozen organic strawberries*
- *1 cup (240 ml) almond milk*
- *1 tablespoon (8 g) maca root powder*
- *1 tablespoon (6 g) goji berries*
- *1 to 2 teaspoons royal jelly and raw honey mix*
- *1 banana*

Directions:

> Blend the ingredients together in a blender.

> Depending on your desired thickness, use more or less milk to get the consistency you like. Serve immediately.

Iron Woman Smoothie

Whether you are iron deficient or looking for an energy boost to power you through your upcoming race, this richly spiced smoothie will help. Royal jelly increases oxygen uptake, boosting endurance and stamina. Royal jelly also helps with anemia, which women are twice as likely to suffer from than men.

Yield: 1 or 2 servings

Ingredients:

- 1 cup (240 ml) almond milk
- 1 teaspoon blackstrap molasses (more or less to taste—it can be quite strong!)
- ¼ teaspoon ground cinnamon
- ¼ teaspoon ground ginger
- ½ teaspoon pure vanilla extract
- 1 frozen banana
- 1 to 2 teaspoons royal jelly and raw honey mix
- A few ice cubes

Directions:

> Blend the ingredients together in a blender.
> Depending upon your desired thickness, use more or less milk to get the consistency you like. Serve immediately.

Bee Venom

A Healing Sting

Only around two in every 1,000 people are allergic to a honeybee sting (in comparison, peanut allergies are about five times more common). So it's unlucky that I am one of those unfortunate few. People are always shocked when they hear that I'm highly allergic to bees, given my passion and profession! But my allergy means that I'm always especially calm and careful when I'm working with my four-winged fuzzy friends. And it has fortuitously introduced me to the interesting world of bee venom therapy.

Did you know that many people use bee venom as a healing tool, helping them treat migraines, arthritis, acne, Lyme disease, multiple sclerosis (MS), and more? Every year, more and more research seems to be emerging on the ability of bee venom to be used for health and healing.

What Is bee venom?

Bees sting only in defense, when they sense an imminent danger to themselves or their hive. Only female honeybees have stingers; scientists believe that this is because the stinger evolved from the bee's ovipositor (the organ through which a female insect deposits eggs). The stingers are very well designed for their mission; barbs on the stinger anchor it to the target. As the bee flies away, the stinger and attached venom sac are torn from the bee's body and left behind, where the venom sac continues to pump venom even after the bee is long gone. (This is why it's important to remove the stinger immediately after a sting; studies have shown that leaving the stinger in even for just eight seconds can increase the size of a local reaction by 30 percent.) Each sting can release up to 0.1 milligram of bee venom, scientifically known as apitoxin, into the victim (less toxin is released if the stinger is removed before the venom sac has completely emptied).

Bee venom is colorless and odorless. It is 88 percent water and slightly acidic. When it is exposed to air, it dries and turns into a white powder. Aside from water, the principal component of bee venom (50 percent of the venom's dry weight) is the protein melittin—the cause of much of the pain of the bee's sting. Melittin destroys blood cells and causes blood vessels to dilate; this is why some people experience a drop in blood pressure after a bee sting. As discussed later, melittin has also been found to be strongly antimicrobial. Other ingredients in bee venom are phospholipase A2, another protein that destroys cell membranes at the site of the sting; histamine, which causes an allergic response within the body; apamin, a neurotoxin; and hyaluronidase, which dilates the capillaries, causing the inflammation to spread.

Once your body is stung, your immune system starts making antibodies to defend against the proteins in the venom. In rare cases (about two in every 1,000 people, including me) the body mounts a severe overreaction to the bee venom, and symptoms are experienced in areas other than the site of the sting, including itchiness, swelling, dizziness, a drop in blood pressure, and shortness of breath. This type of systemic reaction requires immediate medical attention. However, the vast majority of bee stings are not a major medical issue. A local reaction at the site of a bee sting is normal, and even a large local reaction does not indicate an allergy to bee venom. Antihistamines can help treat the itching and pain of a sting, and ice can help considerably with the swelling.

Since ancient times, smoke has been used to calm bees when a beekeeper is working with her hives. Using smoke masks the bees' alarm pheromones and decreases aggressive and defensive behavior significantly, greatly lessening the beekeeper's chances of a sting—and accompanying exposure to bee venom. It is thought that the bees sense the smoke to be a nearby fire, causing them to start filling their bellies with honey in case the hive needs to relocate. This takes their attention away from the beekeeper, giving her a chance to do her work.

How Do Beekeepers Harvest bee venom?

Traditionally and historically, bee venom has been administered with live bees, stimulating them to sting the patient in the area in need of healing. Because this practice is a bit of a challenge for those not familiar with working with live bees, in the 1920s bee venom therapists began preparing extracts by crushing whole bees and preparing an injectable solution. Today, a pure bee venom solution is available—and bees don't have to be harmed in the process. The venom is collected by placing a glass plate with a small electrical charge just outside a honeybee hive. The charge alarms the bees, and the hive's guard bees sting the plate. Because the stinger can't sink into the glass, the stinger remains intact and the bee survives. After 30 minutes the plate is removed from the hive. When it's dry, the venom is scraped off the plate and stored in a clean glass jar. It takes 1 million stings to collect just 1 gram (0.04 ounce) of dried venom!

An apitherapist applies a sting to a patient. Some skilled professionals can do this without harming the bee in the process.

Q&A

1. How many people are allergic to bee stings?
a) 2 in every 10 people
b) 2 in every 10,000
c) 2 in every 1,000

2. How many venomous species are there in the world?
a) 10,000
b) 100,000
c) 1,000,000

1.ANSWER: C / 2. ANSWER B

The Gentle and Misunderstood Honeybee

If you're afraid of creatures with fangs or a stinger, you're not alone. Evolution has equipped us with a healthy fear of venomous stings, and for good reason! There are around 100,000 venomous species, and many of them could cause you serious harm. But unless you're a flower, most bees are uninterested in your presence. A honeybee dies after she stings (the bee can't pull the barbed stinger out, and the bee's body is pulled apart quite gruesomely as she flies away), so using the stinger is truly a last resort for a frightened or threatened bee. Most stings happen by accident when a person sits or steps on a bee, or crushes one accidentally. Only a handful of workers (18 to 21 days old) are assigned the role of "guard bee," and even these bees would rather not sting.

Many people mistake the gentle honeybee for other, less gentle stinging insects. I often get calls for "bee removal" and arrive to find a wasp or hornets' nest, with no bees in sight. Unlike bees, wasps such as yellow jackets, paper wasps, and hornets are naturally more aggressive, and because they have stingers with no barbs, they can sting multiple times. Yellow jacket nests are almost always built in holes in the ground. Hornets and paper wasps build aerial nests made of chewed wood pulp. These are the nests that hang from tree limbs out in the open, and are often wrongly depicted as "bees nests" in children's books.

Bumblebees can also sting more than once, but are not naturally aggressive. They are fat and fuzzy (the furry bodies making them good pollinators, like honeybees), and they live in small nests (40 to 500 individuals) in or near the ground.

Ouch! The bee's stinger is still stuck in the victim's skin. If he scrapes the stinger out quickly, his body's reaction will be less severe.

A Bee or Not-a-Bee?
That Is the Question

Bumblebee

Not a bee! (Paper Wasp)

Honeybee

Not a bee! (Baldface Hornet)

Not a bee! (Yellow Jacket)

Bee Venom

How Has bee venom Been Used Throughout History?

Documented use of bee venom as medicine is thousands of years old. Bee venom therapy (BVT) has been used in China for more than 2,000 years, and apipuncture (*api* meaning "bee") is believed to have been the first form of acupuncture.

In ancient Greece, Hippocrates, the "Father of Modern Medicine" (469–399 BCE), used bee venom, calling it "a very mysterious remedy" and referring to it in his books on medicine. The famous physician Galen (129–216 CE) also wrote about the uses of bee venom. In the Middle Ages, Charlemagne (742–814 CE) was treated with bee stings.

The first scientific paper on the use of bee venom for rheumatic diseases was published in 1859 by Dr. Desjardins of France, in *Medical Bee Journal*, where he described his successful use of bee venom for all kinds of rheumatic diseases. This was followed in 1879 by Dr. Phillip Terc of Austria, who became interested in bee venom when he was accidentally stung by bees and found it immensely helpful with his arthritis. He started using bee stings in his practice, and throughout his medical career he documented the treatment of more than 500 rheumatic patients with BVT.

Here in the United States, the history of bee venom therapy is almost a century old. A Vermont beekeeper, Charles Mraz (1905–1999), suffered from a severe case of rheumatic fever at age twenty-eight. Mraz's pain was so intense that he said he "actually wanted to cut [his] leg off to have one minute of relief." After healing himself and subsequently many others with bee venom therapy, he became recognized as the pioneer of BVT in the United States and helped found the American Apitherapy Society. Throughout his long career, Mraz documented remarkable results using bee venom to treat a variety of inflammatory conditions, including rheumatic fever, arthritis, and MS.

bee venom
for Nutrition, Health, and Wellness

Bee venom therapy is far from widespread, perhaps because it requires a significant amount of effort and pain to heal yourself through repeated stings. In a culture where we are always searching for a "magic pill" that can alleviate the discomfort associated with various conditions, it is not surprising that the application of stinging insects has not taken hold. However, around the world, BVT has a true following, especially with acute conditions that are associated with inflammation, such as arthritis, MS, and other autoimmune diseases.

It may seem funny that bee stings, which cause (temporary) pain and inflammation, are used to help alleviate inflammatory conditions, but it makes sense when you look at the science behind the sting. Bee venom stimulates cortisol production, which is a strong anti-inflammatory agent. Bee stings stimulate blood flow at the site of the sting, causing a release of endorphins, which can help relieve the discomfort of arthritis.

Bee venom contains melittin, which has well-documented success with suppressing inflammation. Melittin is antimicrobial, and it has been shown to inhibit the bacteria that causes Lyme disease. In fact, a growing number of people are praising BVT for its effectiveness in helping to treat their Lyme symptoms. Melittin's antimicrobial property is perhaps why bee stings have been found effective for acne. Because acne is an infection under the skin, it is perhaps bee venom's antibiotic properties that cause the acne to clear in a way that topical antibacterials can't achieve.

Now that bee venom can be collected in a clean and standardized manner, it can also be applied topically in the form of a cream or an ointment. Bee venom creams are widely available in natural markets and online, and are reportedly used by Kate Middleton and Gwyneth Paltrow as a "natural Botox." Perhaps because I feel I've earned my laugh lines, or perhaps as a result of my bee venom allergy, I have never felt compelled to rub bee venom on my face, but if you try it, I'd love to hear about your experience!

Venom Immunotherapy

Bee venom can also be used to help decrease sensitivity to bee stings in allergic individuals, like me. To help desensitize my body to bee venom, I am "stung" intentionally once a week in a controlled environment. My weekly treatment started with the tiniest of stings, and slowly built up to a venom dose that's equivalent to two full stings at once. After several months of stings, my body stopped reacting strongly to the venom. Immunotherapy is highly effective and can reduce the body's reactions to stings for up to ten years, even after treatment stops.

Robin and Juan's Story

A few years ago, Mr. Benevolent Bee and I visited the lovely town of Puerto Escondido, a beautiful vacation spot on the very southernmost tip of Mexico. We were fortunate enough to stay at the beautiful beachfront Hotel Santa Fe, and spend some time with Robin, the hotel's owner. Robin injured his right knee in a skiing accident in his teenage years, and it had never fully healed. Many years later he was walking high in the hills at his coffee farm when he reinjured the joint. While doctors recommended an expensive and invasive surgery, a friend urged him to try bee sting therapy, which he had heard was very effective for inflamed conditions. Willing to try anything to decrease the pain, Robin called the farm's beekeeper, Juan, and asked him to come over and bring some bees to sting his knee. Observing practically immediate results, Juan began stinging Robin weekly, and Robin's knee has been better ever since!

A few years later, bee stings saved Robin again. Robin herniated a disk in his lower back while moving a piece of furniture at home. He visited a chiropractor but was unable to find relief. He called Juan, who came over with some bees and stung Robin several times. The bee venom acted as an anti-inflammatory, and allowed the disk to relocate.

Word spread, and these days, the Hotel Santa Fe serves as an informal apitherapy treatment center. People of all ages line up twice a week to receive stings from Juan, treating all manner of ailments. Juan brings a frame of bees from a nearby hive to an open-air pavilion where he applies anywhere from one to more than twenty stings to each person in line. Patients with migraines are stung on their heads; those with arthritis, back pain, or other injuries are stung at the site of their pain and inflammation. After holding the bee in his fingers and administering the sting, Juan uses tweezers to lift the bee off of the patient in a way that enables the bee's stinger to stay intact in the bee, so the bee can survive after stinging.

Photo by Tony Richards, visitapuerto.com

Juan tends to a frame of bees, calming them with smoke before choosing one to use in administering a treatment.

Photo by Tony Richards, visitapuerto.com

Robin Cleaver stands ready to receive a dose of apitherapy.

In Robin's words, "Bees changed my life!"

!

This information is for educational purposes only.

It is not intended to replace the advice of a physician or medical practitioner. Please see your health-care provider before beginning any new health program. Do not engage in apitherapy treatment if you are or may be allergic to bees.

Recipes

Because bee venom is a bit different than the other products of the honeybee hive in that it's not widely available or recommended for casual use, the recipes in this chapter are different, too. While not actually including venom, each recipe contains a "sting" of sorts, as well as health-promoting ingredients from other hive-derived products.

Spring Allergy Relief: Stinging Nettle Tea with Honey

Have you ever brushed against stinging nettle leaves? You'd know it if you had. The leaves have fine hairs that contain a number of chemicals that cause redness, stinging, and itching on the skin that lasts for about four hours. But did you know that in addition to their sting, nettle leaves also contain natural antihistamines and anti-inflammatories? Stinging nettle has been used medicinally for ages, dating back to ancient Greece. Combined with local honey, nettle tea is a great daily tonic for people suffering from spring allergies, relieving itchy and watery eyes, sneezing, and runny nose. The emerald green tea is also highly nutritious, as it contains iron, calcium, vitamin C, and tons of other nutrients.

If you harvest the leaves yourself, pick the leaves before the flowers form, and only harvest the nettle tops (the top four leaves). Plants that are less than knee-high will have the softest leaves. Discard the stems. Be sure to wear strong gloves and protective clothes!

Yield: 4 cups (940 ml)

Ingredients:

- *1 cup (30 g) fresh nettle leaves or ¼ cup (7.3 g) dried nettle leaves*
- *4 cups (940 ml) water*
- *Slice of lemon*
- *Raw honey, to taste*

Directions:

› Always use gloves when working with fresh leaves!
› Place the nettle leaves in a pot, add the water, and simmer over medium to medium-low heat for 15 minutes.
› The water should take on the beautiful green color of the nettle leaves.
› Remove from the heat. Squeeze the lemon slice into the nettle liquid.
› Add raw honey and stir.
› This tea is delicious hot or iced!

The Benevolent Bee's Bee Sting Cocktail

This sweet and spicy, slightly sour drink hits all the bases on your palate and will warm you right up! Plus, with the healthful ginger and honey combination, and vitamin C from the citrus, it's immune stimulating and nutritious to boot. So drink to your health!

Yield: ½ cup (120 ml) simple syrup and 1 cocktail

Ingredients:

Ginger Honey Simple Syrup
- *2 tablespoons (16 g) grated fresh ginger*
- *1 teaspoon grated organic orange zest*
- *1 cup (240 ml) water*
- *½ cup (170 g) honey*

Cocktail
- *2 ounces (60 ml) bourbon*
- *1 ounce (30 ml) ginger syrup (see above)*
- *Juice of 1 lemon, freshly squeezed*
- *Ice*
- *Twist of orange or lemon rind, for garnish*

Directions:

> To make the simple syrup, combine the ginger, orange zest, and water in a small saucepan.

> Simmer for 15 minutes over medium to medium-low heat, stirring occasionally, until the liquid is reduced by about half.

> Remove from the heat and let cool a bit before straining through a coffee filter into a glass jar.

> Stir in the honey to taste (adding the honey last after the mixture has cooled a bit preserves the enzymes and bacteria in the raw honey).

> Store in the refrigerator for up to 2 weeks.

> To make a cocktail, chill a cocktail glass in the refrigerator. In a shaker, combine the cocktail ingredients with ice and shake vigorously. Strain into your glass and garnish with an orange twist.

Ginger and Honey
A Powerful Combination

Contributed by herbalist Brittany Wood Nickerson, Thyme Herbal

Ginger root is a powerful medicinal herb, long respected for its diverse application in the home pharmacy and beyond. Ginger's "sting" is warming and stimulates circulation to the digestive track as well as to peripheral limbs. It is good for poor circulation and for people who often feel cold. It is also excellent for the digestive system, improving the breakdown and absorption of nutrients, priming the appetite, and improving metabolic processes. It can be used to ease digestive symptoms, including gas, bloating, cramping, and nausea.

Moreover, it is a powerful antimicrobial, internally effective against foodborne illness and common cold and flu viruses. It can also be used topically to kill bacteria and fight infection. Combine it with honey, also a potent antimicrobial, and you have a powerful combination! The sweet flavor of honey complements the warmth of ginger, soothing and softening the spice of the ginger.

Beeswax

A Clean and Golden Light

One of my favorite things about making candles is the smell it gives the house. Every visitor to our home remarks on this smell—and most are polite enough *not* to remark on the wax drippings on the floor, the product of what we call "waxidents."

With its golden glow and honeyed scent, beeswax has been used since ancient times as a source of light in darkness. Are the candles on your table made of beeswax? If not, they are likely made of paraffin, which is a petroleum by-product. Paraffin candles release black smoke when they burn, polluting indoor air. But when beeswax burns, it emits negative ions, improving the air you're breathing by capturing and neutralizing dust, odors, mold, bacteria, and other toxins.

Not only does beeswax smell good, but it's dripless and long burning, too. In fact, with the right wick, beeswax can burn five times longer than other waxes.

What Is beeswax?

There are nearly 20,000 different species of bees, but only honeybees make beeswax, using it to create the structural framework of their hive—the honeycomb. It is where nectar and pollen are stored, where honey is produced, and where eggs develop into larvae and then baby bees. Amazingly, when beeswax is produced by the bees, it is pure white and has no distinct odor. It is only through its life of use in the hive that it takes on its distinct golden color and delicious scent.

Beeswax is produced by worker bees that are 12 to 18 days old. Each worker bee has eight wax-secreting glands under its abdomen, and each gland secretes a sliver of warm wax in the form of a small scale, the size of a pinhead. Other worker bees collect the wax from the wax-producing workers with their mandibles and soften it in their mouths.

The creation of beeswax is an expensive process for the bee—8 pounds (3.6 kg) of honey are required for every 1 pound (454 g) of wax created! This is why bees use the amazingly efficient hex-agonal shape in their hive design. The hexagons in the honeycomb fit together perfectly, without leaving any space. Bees are incredibly exact in their architecture. Each of the six hexagonal cell walls meet at an angle of precisely 120 degrees, and each wall is exactly the same width. The efficiency of this design has been commented on for centuries. The Greek mathematician Pappus of Alexandria (c. 290–c. 350 CE) stated that bees "possessed a divine sense of symmetry." Charles Darwin (1809–1882) called the honeycomb "absolutely perfect in economizing labor and wax."

How Do Beekeepers Harvest beeswax?

For the beekeeper, beeswax is most often a by-product of honey processing. Bees seal each honey-filled cell with a beeswax cap. To extract the honey, a beekeeper either uses a hot knife to slice off the wax cappings, exposing the honey inside, or she simply crushes the whole comb over a fine-mesh strainer. The wax cappings and crushed wax comb remain on top of the mesh while the honey drips through, ready to be bottled.

Turning Honeycomb into Useable Beeswax

1. Using a dedicated melting pot, melt the raw wax comb (A) in boiling water (always making sure to use more water than wax).

2. Let the wax and water cool entirely. Any residual honey will have dissolved in the water, and particulates in the wax will have settled at the bottom of the wax cake (B).

3. Scrape the particulate layer off of the bottom of the wax cake.

4. Melt the wax again, this time using a double boiler.

5. Strain the wax into a silicone baking pan (C). Some good materials to use as strainers include an old T-shirt, nylon mesh paint strainer, and doubled cheesecloth.

6. Repeat step 5 as necessary until your wax block is clean (D)!

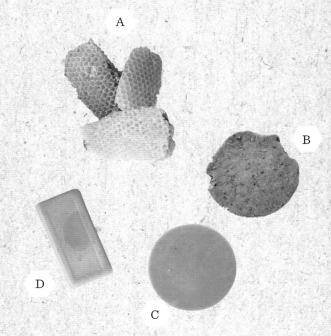

Q&A

1. Wax is produced by bees that are 12 to 18 days old in the form of scales, secreted from the bee's abdomen. One pound (454 g) of beeswax is made from how many wax scales?
a) 8,000
b) 80,000
c) 800,000

2. Honeybees must consume how many ounces of honey in order to produce 1 ounce (28 g) of beeswax?
a) 2
b) 4
c) 8

1. ANSWER: C / 2. ANSWER: C

How Has beeswax Been Used Throughout History?

Beeswax is one of the most useful products in the world. It's edible, moldable, waterproof, smells amazing, burns bright and long . . . and so much else, too. Beeswax has been used by humans for a long time, and in an amazing variety of ways.

- One of the earliest recorded uses of beeswax was in ancient dentistry; in fact, the earliest evidence of dentistry was uncovered in Slovenia when scientists found a 6,500-year-old cracked tooth that had been repaired with a beeswax filling.

- Almost as ancient, beeswax was used as a sealant on the inner surface of vessels as early as 3700 BCE, in Neolithic settlements in Bavaria.

- In Egypt, beeswax models of divinities, humans, and animals have been found from as early as 2830 BCE. And in Australia, Aboriginal beeswax paintings have been found on rock surfaces dating back to 2000 BCE.

- Beeswax candles have been in use for at least 3,500 years; evidence of candles can be found in tomb paintings from early Egypt, and an early candlestick from around 1600 BCE was found in the Greek city of Crete.

- Homer described one use for beeswax in the *Odyssey* (written c. 700 BCE); Ulysses used it to help his sailors plug their ears to the song of the Sirens.

- The Greek physician Hippocrates (460–375 BCE) used beeswax as a healing tool, specifically recommending its application for boils and abscesses.

- In ancient Rome, Pliny the Younger (100 CE) described the creation of a lantern of sorts by dipping the dried pith of the rush plant into beeswax. Beeswax lights were widely used during medieval times. And because it was valued so highly, beeswax was used as currency to pay land taxes.

A female figure is outlined in beeswax on top of a painting dating from between 1762 and 1814 in Aboriginal Australia.

beeswax
for Nutrition, Health, and Wellness

Today, beeswax is used in a number of applications. There's a good chance that you use it daily, whether you know it or not! It may be in your mustache wax or dental floss, on your cheese or in your candy, in your floor polish or on your snowboard. Useful, edible, and nontoxic, beeswax is everywhere—and for good reason!

Beeswax is clean and nontoxic. When combined with herbal oils, beeswax ensures a smooth, even, and drip-free distribution of the oils on skin, making it perfect for use in ointments and healing creams. The cosmetics industry is the biggest consumer of beeswax today, using it for everything from hand cream to lipstick, hair products, and mascara. The second-biggest user of beeswax is the pharmaceutical industry. Because it is edible, beeswax is often used as a coating for medicines that are most effective when absorbed by the lower gut. The coating helps the pills travel through the stomach before being digested. It's also used in medical lubricants, pharmaceutical-grade creams, and more.

When beeswax is produced by the bees, it is completely white and odorless. Its eventual golden color and honeyed scent are due to the incorporation of small amounts of pollen, propolis, and honey in the wax, resulting from its use in the hive. Because of these elements, beeswax also carries some of the antibacterial, antifungal, and antioxidant properties of honey, making it an especially good choice for use in health and healing.

Projects

Beeswax is a beautiful, sweet-smelling, and useful substance. And yet many beekeepers do not use the beeswax from their hives! This is perhaps because it takes time and dedication to process the raw wax into a clean and usable form. But if you have the time, or the luxury of starting with clean and usable wax on hand, working with beeswax can be simple, practical, and enjoyable for all ages.

What Is "Bloom"?

Over time, a white chalky powder will accumulate on all 100 percent beeswax candles. This is caused by sugars in the candle making their way to the surface—it's called "bloom," and it's a completely natural occurrence and a sign that your beeswax is pure. It doesn't affect the quality or burning of the candle. Some people prefer the antique or textured appearance, but if you want your candles to look shiny, you can just wipe them gently with a cloth that is either dry or dipped in a little oil, rinse them in water, or heat the candle lightly with a hair dryer.

Simple Container Candle

A vintage glass, a Mason jar, a votive container, an orange peel—any container can become a candle with this simple method.

Yield: 1 candle

Ingredients:

- *Beeswax*
- *Vessel of your choice (glass cup, tea-light cup, votive container, Mason jar)*
- *Tabbed wick*
- *Double boiler*
- *Chopsticks*
- *Scissors*

Directions:

> Melt your beeswax in a double boiler. Pour a tiny bit of wax into the bottom of your vessel.

> Place a tabbed wick in the middle of the vessel, in the melted wax. It should feel like the cooling wax is grabbing the tab and holding it in place.

> Let it cool a bit, then secure your wick in the middle of the vessel using chopsticks as shown in the image.

> Carefully pour wax into the vessel.

> Leave a bit of room at the top in case any cracks or holes develop as the candle cools; you can melt more wax to fill and cover any imperfections.

> Cracks develop when the wax cools too quickly. It's good to pour your candles in a warm room, and on a surface that holds heat.

> If you're having trouble with cracks, try insulating your container with a towel.

> Trim the wick with scissors.

Floating Acorn Cap Candle

Fall often finds me and my kids on our knees underneath an oak tree, hunting with the squirrels. I have too often opened up the washing machine to discover that I forgot to empty our pockets before putting our clothes in the wash. But it's all worth it. These very sweet floating candles are great for celebrations or gatherings. They don't have the longest burn time (only 10 to 20 minutes, depending on the size of your acorn cap), but their whimsy is unparalleled. I particularly like using them to decorate our Thanksgiving table.

Yield: Varies

Ingredients:

- *Acorn caps*
- *Dry rice*
- *Bowl*
- *Beeswax*
- *Double boiler*
- *Tabbed wick*
- *Scissors*
- *Decorative bowl filled with water*

Directions:

› Make a nest for your acorn caps using a bowl of dry rice. The rice will help keep the caps upright and level while you're pouring in the wax.

› Melt the beeswax in a double boiler.

› Place a tabbed wick in the middle of each cap, and carefully pour the wax into each tiny vessel.

› It takes practice and a steady hand, but after a few caps, you'll get the hang of it and be able to fill each cap perfectly!

› Trim the wicks with scissors.

› Float the candles in a decorative bowl filled with water.

Beeswax

Rolled Candles

Rolled candles are perhaps the easiest to make, as they don't require a stove. I love to make these candles with kids of all ages; they come together quickly, burn beautifully, and can be made in a number of shapes and sizes, from thin birthday candles to wide pillars. Tie a pair of rolled tapers with a length of ribbon for a lovely gift handmade in minutes.

Yield: Two 8-inch (20 cm) candles

Ingredients:

- *1 beeswax sheet (they are typically sold in sheets of 8 × 16 inches [20 x 40 cm])*
- *Scissors*
- *2 pieces of wick, each 9 inches (23 cm) long*

Directions:

> Cut your beeswax sheet in half to make two 8 × 8- inch (20 × 20 cm) square sheets.

> Lay the wick along the rough edge of one sheet, with one end of the wick flush with the side of the sheet and the other hanging off the edge (this will be the piece of the wick that you light).

> Begin rolling the candle evenly.

> Try to roll it up tightly, but not so tight that you crush the wax.

> When you reach the end, gently smooth the edge of the wax into the candle.

> Repeat with the other sheet.

> Trim the wicks.

Variations:

> You can make rolled candles any height and thickness. It's easy to do!

> First, decide on the height of your desired candle. Since the beeswax sheets are 8 × 16 inches (20 × 40 cm), it's nice to choose a size that is a divisor of 8; for instance, a birthday candle could be 2 inches (5 cm) tall, 8 inches (20 cm) is a nice size for a taper candle, and 4 inches a good height for a pillar.

> Cut the wick to 1 inch (2.5 cm) longer than the height of your desired candle.

> Gently cut the sheet of beeswax lengthwise to the height of your desired candle.

> As you roll the candle, pay close attention to thickness. When you reach your desired width, simply cut the sheet and gently press the cut edge into the candle, smoothing the rough edge. If you'd like a thick pillar, you can add a sheet of beeswax when the first one runs out, and continue rolling until your candle is nice and fat.

Homemade Molds

Once you start making your own candle molds, you'll be seeing the world with different eyes. Anything and everything becomes a potential candle form. That piece of fruit might make an interesting candle! And what about an egg? A child's toy, a jar, a stick . . . it's really fun to get creative!

Yield: 1 mold

Materials:

- *Piece of wood (optional)*
- *A long screw or nail (optional)*
- *Screwdriver or hammer (optional)*
- *Glue (optional)*
- *Item for molding*
- *Plastic container wide enough to completely fit your item without touching (like a quart-size deli container)*
- *Silicone mold-making material (we use OOMOO 30 Silicone Rubber from Smooth-On)*
- *Scissors*
- *Wicking needle*
- *Wick*
- *Beeswax*
- *Double boiler*
- *Chopsticks*

Directions:

› Screw through the wood and into the bottom of your item (the side that will be the bottom of the candle), leaving a gap of a few inches (5 to 7.5 cm) between the wood and your item.

› Place your item inside the plastic container with the wood resting on the container's top, making sure that your item does not touch the bottom or sides of the container. If it does touch, find a bigger container.

› If your item is made of a material you cannot screw into (such as the glass container in these photos), glue the item to the inside of the container.

› Mix the silicone mold-making material as described on the packaging (A), and pour it around your item in the container (B). Fill the container up to the bottom of your item, leaving the very end of your item visible above the mold material. Let the mold harden completely.

› Pop the now-hard mold out of the container (C).

› Cut carefully down one side of the mold with scissors, cutting only as much as you need to release the item inside (D, E).

› Using a wicking needle, pull the wick through the center of the mold.

› Melt the beeswax in a double boiler.

› Place the mold back into the plastic container, and fill the cavity with the melted beeswax, holding the wick in the center of the mold with the chopsticks.

› Let the wax cool, and then remove the candle from the mold, pulling more wick through into the mold cavity, getting you ready for another pour!

Which wick type is best for these candles? These tea-lights were each made with a different wick type for a burn-time test.

Choosing the Right Wick

With the right wick, beeswax candles will burn up to five times longer than candles made from paraffin wax, and are completely dripless. Using too small a wick will lead to "tunneling" or will cause the flame to drown out. Using too large a wick will make the candle burn too fast, causing drips and reducing the candle's burn time.

There are a lot of options when choosing a wick (there are more than 300 different wick types!), but don't let it overwhelm you. With a few deep breaths, a bit of research, and some trial and error, your candles will burn perfectly (and even when you get it wrong, they'll still be beautiful!).

The reference chart on the right will help you choose the right size wick for your candle project. For all candles, choose a 100 percent cotton, square braid wick.

Recommended Wick Size	Candle Diameter
4/0	Tea light
3/0	Standard votive ($1\frac{1}{2}$" [3.8 cm] diameter)
2/0	Standard taper ($\frac{7}{8}$" [2.2 cm])
#1	1"–$1\frac{1}{2}$" (2.5–3.8 cm)
#2	$1\frac{1}{2}$"–2" (3.8–5 cm)
#3	2"–$2\frac{1}{2}$" (5–6.4 cm)
#4	$2\frac{1}{2}$"–$2\frac{3}{4}$" (6.4–7 cm)
#6	$2\frac{3}{4}$"–$3\frac{1}{4}$" (7–8.3 cm)
#7	$3\frac{1}{4}$"–$3\frac{1}{2}$" (8.3–8.9 cm)
#8	$3\frac{1}{2}$"–4" (8.9–10.2 cm)

Hand-Dipped Birthday Candles

Making these candles is a lovely ritual that we repeat every time there's a birthday in the family. Suitable for ages three and above, this project is fun, easy, and lends the birthday cake an extra-special handmade touch. They're also wonderful birthday (or holiday) gifts.

Yield: 1 candle per dipped wick

Materials:

- Wick
- Scissors
- Beeswax
- Double boiler
- Pencil, skewer, or other straight tool
- Metal washers or nuts
- Deep pot of cold water
- Parchment or waxed paper

Directions:

> Cut the wicks a few inches longer than your desired candle length. Melt the wax in the double boiler.

> Tie a wick onto a pencil or skewer, and then tie a metal washer onto the bottom of your wick.

> Slowly lower the wick into the melted wax, dipping it to the depth you would like your candles to be.

> Pull it out fully, and then dip into the cold water.

> Alternate dips between wax and water, until your candle is the desired width.

> After the first one or two dips, straighten the wick if necessary.

> Birthday candles will take only a few dips; more substantial tapers will require more.

> A standard taper candleholder is ⅞ inch (2.2 cm) wide; if you are making tapers for a candlestick holder, you will want to keep this diameter in mind.

> Lay the candles on a sheet of parchment or waxed paper to harden, or simply suspend the skewer between two chairs.

> Cut off the washers. Trim the top wick to ¼ inch (6 mm). Enjoy!

Honeypot Luminaries

Simple and gorgeous, these lanterns make lovely gifts and are really fun to craft with kids. Each one is unique! You can make the project more advanced by including leaves or petals or by carving designs into the finished project.

Yield: 1 luminary

Materials:

- Beeswax
- Double boiler large enough to dip ballon in
- Balloon
- Parchment paper
- Dried leaves, dried flowers, or tissue paper designs (optional)
- Electric skillet (optional)
- Carving tools such as those sold at Halloween for pumpkin carving (optional)
- Beeswax or battery-operated tea light

Directions:

> Melt beeswax in the double boiler.

> Fill the balloon with cold water.

> The size and shape of the balloon when full will determine the shape of your lantern.

> Carefully dip the balloon partway into the melted wax, and then pull it back out.

> Set it down on the parchment paper to create a small flat spot on the bottom of the lantern so the finished product will be able to rest sturdily.

> Repeat the dipping and resting process several more times until the walls of the luminary have reached your desired thickness.

> If you are adding leaves or petals, you can dip them into the wax and then carefully press them into place on the still warm luminary, and then dip one final time over the design to seal it in place.

> When your luminary is complete and the wax is cool, hold the balloon over the sink and pop it carefully with a small pin, letting the water drain slowly.

> If your balloon doesn't sit well, you can melt the bottom a bit more using an electric skillet.

> If desired, carve the wax lightly with the carving tools.

> If your luminary is big enough, you can light it with a tea light.

> If it is small and you are concerned about the flame of the light melting the wax of the luminary, you can use a battery-operated tea light instead to create your golden glow.

Herbal Salves

So easy and fun to make, salves can be crafted to fit your family's needs by using different infused oils. These instructions will help you make a basic moisturizing hand and body salve that we make and use all the time. This basic recipe can be modified as desired. See below for some suggestions of different herbs to try, depending on your needs.

The addition of essential oils can make your salve smell lovely and can provide additional benefits. Just to name a few, lavender calms and soothes; peppermint is cooling and can ease headaches; tea tree is antibacterial; rosemary can help prolong the shelf life of your salve; and eucalyptus is wonderful in salves you plan to rub on the chest during times of winter congestion.

Vitamin E oil is antioxidant, which is good for your skin, and can also help preserve the shelf life of your salve.

Yield: Makes about 9 ounces (270 ml)

Materials:

- *Digital scale*
- *8 ounces oil(s) (see "Good Oils for Balms and Salves" on page 143)*
- *1½ ounces (43 g) beeswax*
- *1 teaspoon vitamin E oil (optional)*
- *Double boiler*
- *About 60 drops essential oils as desired (depending on the strength)*
- *Containers with lids*

Directions:

› Use a digital scale to measure the oil and beeswax.

› Heat the oil, beeswax, and vitamin E oil in a double boiler over low heat, stirring occasionally, until everything is melted.

› Remove from the heat and add essential oils. Start with a small amount, and add more as desired.

› Pour into containers.

› Store covered in a cool, dry place.

Variations:

› **Opening winter chest rub:** Great for cold season. Rub on chests and the soles of feet before bedtime. Use a mix of eucalyptus, peppermint, and rosemary essential oils.

› **Nourishing hand and body balm:** Wonderful for daily use; smooths and nourishes skin. Also a great diaper salve! Use calendula, plantain, and comfrey essential oils.

› **Soothing headache rub:** Rub a little on your temples and the back of your neck when a headache strikes. Use lavender and peppermint essential oils.

› **Sleepy time massage balm:** Especially nice for children. A loving body rub always helps our kids find their way to sleep. The essential oils in this balm help with relaxation. Use a mix of lavender, chamomile, and marjoram essential oils.

› **Propolis wound care balm:** Add ½ tablespoon propolis powder or extract, and use rosemary, lavender, and tea tree essential oils.

The Only Lip Balm You'll Ever Need

Lip balm is simple and inexpensive to make yourself! Make this recipe your own, and you'll never need to buy lip balm again. You can adjust the recipe to your own preferences. If you want a harder lip balm, add a bit more beeswax. Softer, add less. You can make it unscented, or include some peppermint, orange, or other essential oil for the smell and other added benefits. Cocoa and shea butters are some of nature's richest moisturizers. Cocoa butter can smell very chocolaty, which is quite nice when combined with peppermint essential oil. Shea has an earthy smell that not everyone loves, but I don't mind it at all. I think shea combines nicely with orange essential oil. If using tubes for your containers, it's best to get an inexpensive tube holder (filling tray) and a small spatula. The filling tray makes pouring into the tiny tubes much easier! The spatula helps spread the mixture at the top of the tube, scraping away excess.

Yield: about 25 tubes, 0.15 ounces each

Materials:

- 12 tablespoons (180 ml) oils (we use 1 tablespoon [15 ml] castor oil, 8 tablespoons [120 ml] sweet almond oil, and 3 tablespoons [45 ml] olive oil)
- 4 tablespoons (60 g) beeswax
- 4 tablespoons (60 g) cocoa or shea butter
- Digital scale
- Double boiler
- 25 drops essential oil, such as peppermint, sweet orange, or tea tree (optional)
- Containers with lids
- Filling tray and small spatula, if using tubes

Directions:

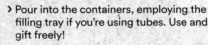

› Measure oils, beeswax, and cocoa butter with the scale. Warm them in a double boiler over low heat, stirring occasionally, until everything is melted.
› Remove from heat and add essential oils. Start with a small amount, and add more as desired.
› Pour into the containers, employing the filling tray if you're using tubes. Use and gift freely!

141

Infused Herbal Oils

Infused oil is easy to make, and useful. Garlic and rosemary infused oil, for instance, can be delicious in a dinner, while other infusions can add the medicinal properties of a plant to the oil used in a lotion or salve.

<u>Yield:</u> 2–3 cups (475–710 ml)

<u>Materials:</u>

- *2–3 cups dried herbs (See "Suggested Herbs for Oil Infusions" at right. It's important that the herbs are dry, as the water in green plants may make the oils go rancid.)*
- *Quart-size (liter-size) Mason jar*
- *Enough oil to cover the herbs by 1–2 inches (8.5–5 cm) (see "Good Oils for Balms and Salves" at right; we most often use extra-virgin olive oil)*
- *Chopstick*
- *Double boiler (optional)*

<u>Directions:</u>

› Place the dried herbs into a clean glass jar.
› Cover the herbs with oil, leaving room at the top of the jar, as the herbs will expand when they soak up the oil.
› Use a chopstick to stir the oil into the herbs, releasing all of the trapped air in the plant material.
› Place a lid on the jar, and put it in a warm and sunny window, agitating the jar daily.
› Gradually, you will notice the oil taking on the color of the herbs. After 3 to 5 weeks, your oil will be ready!
› Strain out the herbs and store the oil in a cool, dark place. It will last for around a year.

<u>Variation:</u>

› If you don't have time to wait a few weeks for your oil to be ready, you can make your infusion overnight on the stove.
› Place your sealed glass jar in a double boiler with water.
› Gently heat the oil on the stove for 1 to 2 hours (make sure the water doesn't go above 140°F [60°C]).
› Turn off the heat and let the oil sit in the warm water bath overnight. In the morning, repeat the warming process for another 1 to 2 hours.
› Remove the jar from the water and let the oil cool before straining and storing as described above.

Good Oils for Balms and Salves

Apricot Oil

Light, gentle, and very moisturizing, this is great for a massage oil for people with sensitive skin. No oily residue!

Castor Oil

Antimicrobial and antifungal. I use a bit of castor oil in almost every balm I make. It's thick, so it's best to dilute it with other oils. Won't clog pores and is good for the skin, especially when used on the face.

Coconut Oil

Very moisturizing, but not great for use with oily skin or on the face. Extremely resistant to rancidity; long shelf life.

Jojoba Oil

Great for all skin types but especially acne-prone skin, as it breaks down excess skin oils. Very rich in vitamin E, with a long shelf life.

Olive Oil

Suitable for all skin types, and great for those with nut or seed allergies. We use it a lot, but it doesn't absorb as well as some other oils.

Sunflower Oil

Excellent skin softener, very nourishing with vitamins A, B, D, and E. Will not clog pores.

Sweet Almond Oil

Softens and conditions the skin. Also rich in vitamin E, with a long shelf life.

Suggested Herbs for Oil Infusions

Arnica

Can help relieve pain from injuries, such as bruises, muscle pain, strains, or swelling.

Calendula

Useful for skin irritation, abrasions, rashes, and cuts. Wonderfully soothing.

Chamomile

Calms skin abrasions, cuts, and scrapes.

Comfrey Leaf

Speeds healing after trauma to muscles or bones; can help with pain and swelling associated with injury.

Ginger

Very warming; can help with sore muscles or as a nice winter rub.

Lavender

Calming, antibacterial; helpful with burns, cuts, and scrapes.

Plantain

Anti-inflammatory; can help treat sunburns, insect bites, poison ivy, rashes, burns, blisters, and cuts.

St. John's Wort

Antibacterial; speeds healing from cuts, scrapes, and bruises; helpful for nerve pain and soothing itch from insect bites.

Yarrow

Helpful for bruises and in cases of swelling and bleeding. Speeds healing of skin wounds.

Postscript

Are Bees in Trouble?

One of the questions that I'm most commonly asked when giving a presentation about bees is, "Are bees in trouble?" There has been a lot of buzz in the media in the past few years about "Colony Collapse Disorder"—the recent dramatic decline in honeybee populations in the United States and worldwide. And for good reason—bees are important! Aside from the simple value of their fascinating and fuzzy existence, they are crucial to our agricultural system and national economy.

In 2006, beekeepers across the country started experiencing a nightmare-like phenomenon: almost all of the bees in seemingly healthy, thriving hives would one day suddenly vanish, leaving only a very lonely queen bee. Termed "Colony Collapse Disorder" (CCD), these disappearing hives occurred throughout the country, causing the demise of about 24 percent of the nation's apiaries that year. This led to significant economic devastation not only to beekeepers but also to the agricultural industry that relies heavily on the bees' pollination services. Scientists and the government agencies started scrambling to figure out what was going on. Theories ranged from pesticide use to climate change, electromagnetic radiation to genetically modified crops, and more. We still don't really know why CCD came on so suddenly, or what, exactly, its causes were. But it does seem that this exact phenomena—of a colony suddenly vanishing without warning, and without a trace left behind—is largely over. No cases of CCD have been reported in the past several years.

CCD might be over, but challenges facing honeybees are not. It turns out that honeybee populations have been in decline since the 1950s—a slower, less dramatic decline than hives mysteriously vanishing with CCD, but just as devastating to beekeepers and bee colonies. Nationwide, bee colonies have decreased by an estimated 50 percent since the 1950s, and beekeepers continue to be plagued by steep winter losses. In fact, 44 percent off all U.S. bee colonies died in the winter of 2015/2016—a devastating figure.

Why? It seems that the cause isn't just one but several factors, all working together to make life tough for honeybees. The 1950s marked the start of commercial agriculture and the rise in "monocrop" farms—vast areas where only one crop is planted, severely limiting pollinators' natural foraging opportunities. Commercial beekeeping rose hand-in-hand with commercial agriculture; because large farms no longer naturally host pollinators, beekeepers are paid to relocate their bees temporarily onto the land when the crop is flowering. This happens with blueberries in Maine, cranberries in Massachusetts, oranges in Florida, almonds in California, and elsewhere across the nation. Life on the road is not natural for bees, and the stress of travel weakens the bees' ability to fight off parasites and disease. There has also been a marked increase in pesticide use during this period of bee decline, including neonicotinoid pesticides that can accumulate in the hive, damaging the bees' nervous systems and hampering their ability to forage and fly.

In late 2016, literally millions of honeybees were poisoned by an aerial pesticide spray of the neurotoxin Naled to fight the Zika virus. The spraying was conducted in South Carolina, where, to date, no Zika-carrying mosquito has ever been found. With warming weather patterns causing a longer "mosquito season" in cities across the nation, and the spread of mosquito-borne illnesses such as Zika and West Nile virus, this type of spraying is likely to become more common. My hope is that beekeepers can work with the federal government and local municipalities on regulations for the use of powerful neurotoxins like Naled in a way that can protect pollinators, human health, and the environment.

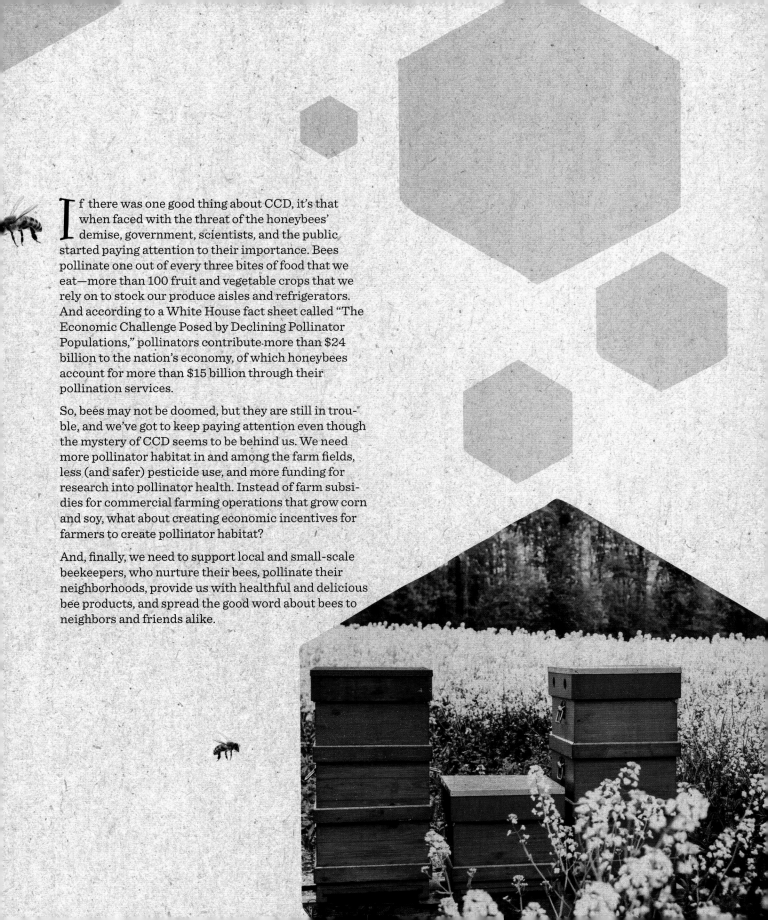

If there was one good thing about CCD, it's that when faced with the threat of the honeybees' demise, government, scientists, and the public started paying attention to their importance. Bees pollinate one out of every three bites of food that we eat—more than 100 fruit and vegetable crops that we rely on to stock our produce aisles and refrigerators. And according to a White House fact sheet called "The Economic Challenge Posed by Declining Pollinator Populations," pollinators contribute more than $24 billion to the nation's economy, of which honeybees account for more than $15 billion through their pollination services.

So, bees may not be doomed, but they are still in trouble, and we've got to keep paying attention even though the mystery of CCD seems to be behind us. We need more pollinator habitat in and among the farm fields, less (and safer) pesticide use, and more funding for research into pollinator health. Instead of farm subsidies for commercial farming operations that grow corn and soy, what about creating economic incentives for farmers to create pollinator habitat?

And, finally, we need to support local and small-scale beekeepers, who nurture their bees, pollinate their neighborhoods, provide us with healthful and delicious bee products, and spread the good word about bees to neighbors and friends alike.

Resources

American Apitherapy Society (AAS)

www.apitherapy.org

The American Apitherapy Society collects and disseminates research, case studies, and personal stories about the use of beehive products for health and healing. Their website is a great source of information for individuals interested in learning more.

American Beekeeping Federation

3525 Piedmont Road

Building 5, Suite 300

Atlanta, GA 30305

www.abfnet.org

The American Beekeeping Federation is a member-based organization that acts on behalf of beekeepers and the beekeeping industry on issues affecting the interests and the economic viability of the various sectors of the industry.

Apimondia

Corso Vittorio Emanuele 101

I-00186 Roma, Italy

www.apimondia.org

Based in Rome, Apimondia is the International Federation of Beekeepers' Associations. It works to promote scientific, ecological, social, and economic apicultural development in all countries.

International Bee Research Association (IBRA)

91 Brinsea Road

Congresbury, Bristol BS49 5JJ UK

www.ibrabee.org.uk

IBRA is the world's longest established apicultural research publishers, promoting the value of bees by providing information on bee science and beekeeping.

Materials

Beeswax

BetterBee
8 Meader Road
Greenwich, NY 12834
www.betterbee.com

Brushy Mountain Bee Farm
814 Old Route 15
New Columbia, PA 17856
www.brushymountainbeefarm.com

Honey

Follow the Honey *
1132 Massachusetts Avenue
Cambridge, MA 02138
www.followthehoney.com

Jars, Tins, and Containers

SKS Bottle
2600 7th Avenue
Watervliet, NY 12189
www.sks-bottle.com

Specialty Bottle
3434 4th Avenue S
Seattle, WA 98134
www.specialtybottle.com

Mold Making

Reynolds Advanced Materials
www.reynoldsam.com

Oils, Butters, and Essential Oils

Brambleberry
2138 Humboldt Street
Bellingham, WA 98225
www.brambleberry.com

Bulk Apothecary
25 Lena Drive
Aurora, OH 44202
www.bulkapothecary.com

Mountain Rose Herbs
PO Box 50220
Eugene, OR 97405
www.mountainroseherbs.com

Organic Herbs

Bulk Apothecary
25 Lena Drive
Aurora, OH 44202
www.bulkapothecary.com

Jean's Greens
1545 Columbia Turnpike
Castleton, NY 12033
www.jeansgreens.com

Mountain Rose Herbs
PO Box 50220
Eugene, OR 97405
www.mountainroseherbs.com

Wholesale Supplies Plus
7820 E. Pleasant Valley Road
Independence, OH 44131
www.wholesalesuppliesplus.com

Wicks

The Candlewic Company
3765 Old Easton Road
Doylestown, PA 18902
www.candlewic.com

*If you're ever in the Boston area, I highly recommend a visit to Follow the Honey in Boston's Harvard Square. All products (including more than 100 varieties of honey) are sourced from small-scale farmers and makers, and all honey sold is raw and treatment free. The store's owner, Mary Canning, is a wealth of information about all things honey, and she has a passion for the products of the hive that transcends retail. Through educational events and support for local beekeepers, Mary and her store aim to "tell the narrative of our collective humanity through honey and bees."

Bibliography

Abdu, M. A.-S. and M. Ali. "Studies on Bee Venom and Its Medical Uses." *International Journal of Advancements in Research & Technology* 1, no. 2 (2012): 69–83.

About Honeybees. Fayetteville, AR: University of Arkansas, Division of Agriculture Cooperative Extension Service, 2016.

Al Somal, N., et al. "Susceptibility of *Helicobacter pylori* to the Antibacterial Activity of Manuka Honey." *Journal of the Royal Society of Medicine* 87, no. 1 (1994): 9–12.

Bee Stings. Washington, DC: United States Department of Agriculture, Agricultural Research Service, 2016.

Bernardini, F., C. Tuniz, A. Coppa, L. Mancini, D. Dreossi, et al. "Beeswax as Dental Filling on a Neolithic Human Tooth." *PLoS ONE* 7, no. 9 (2012): e44904.

Bostock, J., and H. T. Riley, eds. *Pliny the Elder, the Natural History, Book XI. The Various Kinds of Insects.* London: Taylor and Francis, 1855.

Bostock, J., and H. T. Riley, eds. *Pliny the Elder, the Natural History, Book XXII. The Properties of Plants and Fruits.* London: Taylor and Francis, 1855.

Broadhurst, C. Leigh. *Bee Products for Better Health.* Summertown, TN: Books Alive, 2013.

Burdock, G. A. "Review of the Biological Properties and Toxicity of Bee Propolis (Propolis)." *Food and Chemical Toxicology* 36, no. 4 (1998): 347–363.

Castaldo, Stefano, and Francesco Capasso. "Propolis, an Old Remedy Used in Modern Medicine." *Fitoterapia* 73 (2002): S1–S6.

Cohen, H. A., I. Varsano, E. Kahan, E. Sarrell, and Y. Uziel. "Effectiveness of an Herbal Preparation Containing Echinacea, Propolis, and Vitamin C in Preventing Respiratory Tract Infections in Children: A Randomized, Double-Blind, Placebo-Controlled, Multicenter Study." *Archives of Pediatrics and Adolescent Medicine* 158, no. 3 (2004): 217–221.

Crane, Ethel Eva. *The World History of Beekeeping and Honey Hunting.* New York: Routledge, 2013.

Duclos, A. J., C.-T. Lee, and D. A. Shoskes. "Current Treatment Options in the Management of Chronic Prostatitis." *Therapeutics and Clinical Risk Management* 3, no. 4 (2007): 507–512.

Fearnley, J. *Bee Propolis: Natural Healing from the Hive.* London: Souvenir Press, 2001.

Ghisalberti, E. L. "Propolis: A Review." *Bee World* 60 (1979): 59–84.

Grange, J. M., and R. W. Davey. "Antibacterial Properties of Propolis (Bee Glue)." *Journal of the Royal Society of Medicine* 83, no. 3 (1990): 159–160.

Grassberger, M., R. A. Sherman, O. S. Gileva, C. M. H. Kim, and K. Y. Mumcuoglu. *Biotherapy—History, Principles, and Practice: A Practical Guide to the Diagnosis and Treatment of Disease Using Living Organisms.* New York: Springer Science & Business Media, 2013.

Gupta, S. S., O. Singh, P. S. Bhagel, S. Moses, S. Shukla, and R. K. Mathur. "Honey Dressing Versus Silver Sulfadiazene Dressing for Wound Healing in Burn Patients: A Retrospective Study." *Journal of Cutaneous and Aesthetic Surgery* 4, no. 3 (2011): 183–187.

Havenhand, Gloria. *Honey Nature's Golden Healer.* Ontario: Firefly Books, 2011.

Krell, R. "Value-Added Products from Beekeeping." *FAO Agricultural Services Bulletin No. 124.* Rome: Food and Agriculture Organization of the United Nations, 1996.

Kuropatnicki, A. K., E. Szliszka, and W. Krol. "Historical Aspects of Propolis Research in Modern Times." *Evidence-Based Complementary and Alternative Medicine* 2013 (2013): 964149.

More, Daphne. *The Bee Book: The History and Natural History of the Honeybee.* Exeter, UK: David & Charles, 1976.

Paul, I. M., J. Beiler, A. McMonagle, M. L. Shaffer, L. Duda, and C. M. Berlin. "Effect of Honey, Dextromethorphan, and No Treatment on Nocturnal Cough and Sleep Quality for Coughing Children and Their Parents." *Archives of Pediatrics and Adolescent Medicine* 161, no. 12 (2007): 1140–1146.

Peterson, I. "The Honeycomb Conjecture: Proving Mathematically That Honeybee Constructors Are on the Right Track." *Science News* 156 (1999): 60–61.

Pujirahayu, Niken, et al. "Antibacterial Activity of Oil Extract of *Trigona propolis*." *International Journal of Pharmacy and Pharmaceutical Sciences* 7, no. 6 (2014): 419–422.

Acknowledgments

Thank you to my parents and primary school, the Miquon School, for encouraging me from a young age to find and follow my passions in life, and for nurturing in me a deep sense of wonder for the natural world and a willingness to get messy, be creative, use my hands, and have fun whenever possible. Thank you also to my family for enthusiastically testing recipes and craft projects, enduring constantly sticky surfaces, and occasionally napping so that I could do some writing. I'm deeply appreciative of my early bee mentors and colleagues, especially Jean-Claude Bourrut, Matt Smith, and Sadie Brown, and my mentors in herbalism at the Boston School of Herbal Studies. I'm also incredibly grateful to the excellent team at Quarry Books for working with me on this project.

It was such a treat to be able to partner on this project with my cousin, the talented photographer Graham Burns. Graham and I both feel so lucky to have been blessed with family who are friends, and friends who are family.

And finally, and most importantly, thank you to the bees!

About the Author

Stephanie Bruneau is a beekeeper, environmental educator, herbalist, artist, homemaker, and mama. She is the owner of The Benevolent Bee (www.thebenevolentbee.com), a small business selling honey, beeswax candles, herbal body care products, and other handcrafted and hive-derived items. At The Benevolent Bee "Teaching Apiary," located just outside of Philadelphia, Pennsylvania, Stephanie observes, learns, and teaches about bees and bee behavior to students of all ages.

When she is not working with bees or pouring candles, Stephanie can be found stomping in mud puddles with Clara (five, whose second word was *bee*) and Atticus (three, who was under a bee suit up in a tree collecting a honeybee swarm with his mom at four weeks old).

Index

Also Available

The Backyard Beekeeper
978-1-59253-919-2

Beekeeper's Lab
978-1-63159-268-3

Beeswax Alchemy
978-1-59253-979-6

The Beekeeper's Problem Solver
978-1-63159-035-1